I0820277

ANOTHER SORT OF MATHEMATICS

ANOTHER SORT OF MATHEMATICS

Selected Proofs Necessary to Acquire a True Education in Mathematics

J. JACOB TAWNEY

2025

Encounter Books
900 Broadway, Suite 601
New York, NY, 10003

First American edition published in 2025 by Encounter Books, an activity of Encounter for Culture and Education, Inc., a nonprofit, tax-exempt corporation.
Encounter Books website address:
www.encounterbooks.com

Manufactured in the United States and printed on acid-free paper. The paper used in this publication meets the minimum requirements of ANSI/NISO Z39.48–1992 (R 1997) (permanence of paper).

FIRST AMERICAN EDITION

LIBRARY OF CONGRESS CATALOGING-IN-PUBLICATION DATA IS AVAILABLE

Library of Congress CIP data is available under the following ISBN: 978-1-64177-444-4 and LCCN: 2025020017.

To my wife, for whom I have the greatest awe and wonder.

To my children, who have something unique in their souls that can only be satisfied by wondering about mathematics.

In thanksgiving for the minds and hearts that have shaped mine in these matters, especially Michael Austin and James Tanton.

INTRODUCTION

YEARS AGO, I read James V. Schall's book *Another Sort of Learning*. The subtitle was "Selected Contrary Essays on How Finally to Acquire an Education While Still in College or Anywhere Else: Containing Some Belated Advice about How to Employ Your Leisure Time When Ultimate Questions Remain Perplexing in Spite of Your Highest Earned Academic Degree, Together with Sundry Book Lists Nowhere Else in Captivity to Be Found."* Schall's short book sets out to list some things you *should* read but probably were never *required* to read. It is not a curriculum, except maybe one for *life*. Similarly, this book you have in your hands is not a curriculum. It is, however, a sort of list of some things from mathematics you should experience but probably were not ever required to experience. As you will read in the opening chapters, because proof holds pride of place in mathematics, so too is it given pride of place in this book. The reason for this is the effusive nature of truth itself. Upon discovery it de-mands to be shared, and the proof is the medium through which mathe-matical truth is communicated. None of the results in this text are my own, and even the proofs were discovered by greater mathematical minds who have gone before me. They represent, in a small way, some of the best that has been said within the discipline of mathematics.

The twentieth-century philosopher and educator Mortimer Adler (1902–2001) once boldly claimed, "The best education for the best is the best education for all." (He was actually quoting his colleague Robert Maynard Hutchins, a president of the University of Chicago in the mid-century.) His point is, I think, that *all* people deserve to read the best that has been written and spoken throughout the course of human history, and this is as true in mathematics as it is in literature, history, or philosophy. The only difference is that we tend to believe it more in other disciplines. For example, the phrase "I cannot read" elicits such a strong negative reaction that it would rarely be admitted in public. In contrast, the phrase "I am not good at math" is often said with ease and perhaps even a hint of pride. And yet, I maintain that the human mind was made for mathematics as much as it was made for reading.

For those who believe that they are "not good at math" and for those

* I owe a huge debt of gratitude to James Schall for his work, and also for the borderline-plagiarized title of my book. I am under no illusion that what is offered in these pages will come even close to what Father Schall has accomplished in his.

who simply "do not like math," I can only offer a profound apology. If your mathematics education has taught you one or both of these things, then it has let you down, so let me be quite clear. *Mathematics is for everyone.* Everyone can be successful at mathematics, and everyone can certainly train his or her mind to love the beauty of mathematics. Some of the proofs in this book are harder than others, but none require university-level training. I have tried to eliminate unnecessarily complicated vocabulary and symbols so as to concentrate on the ideas themselves. After all, a beautiful proof elicits the *idea* of the theorem it sets out to prove and does not get bogged down in terminology and notation. This does not mean that these proofs can be read without struggle. Elegance and ease are not the same thing. The proofs here are elegant, and it is because of this elegance that they are worth struggling through, or perhaps struggling *with*. Struggling *through* implies that we have to endure the proof. Struggling *with* is a call to plunge into the proof itself, to dwell with the ideas, to have an encounter with the objects, and to let all of this become a part of your memory so that you are, at least in some small way, *formed by it*. You will read as an opener to the first chapter: "There is something unique in the human soul that can only be satisfied by wondering about mathematics." Somewhat audaciously, I actually believe this. And that means, regardless of your background, regardless of any math phobia that you have acquired on your educational journey, this book is for you. Reclaim your mathematical inheritance. Embrace the mathematician within you. Choose to wonder.

PART I
THE NATURE OF MATHEMATICS

CHAPTER 1
WHY MATHEMATICS IS WORTH KNOWING

THERE IS something unique in the human soul that can only be satisfied by wondering about mathematics.[1]

Let that sink in for a moment. Then read it again. How could I possibly make such an audacious claim, let alone use it to launch our mathematical excursion? The answer begins with the premise that the human soul is oriented towards that which is *true*, that which is *good*, and that which is *beautiful*. These three transcendentals are the source and summit of human wonder. Mathematics is worth knowing precisely because it is true, good, and beautiful, and as such it satisfies a unique desire in the human soul.

Mathematics Is Good

We study mathematics because it is worth studying in and of itself, which is to say that we study it because it is good. Maybe more precisely, we study the objects of mathematics because they are good. In mathematics we study the pure forms of the heavens, shapes that are so perfect that we have never seen them and numbers that are so large that we have trouble finding physical counterparts, and we are constantly moving in and out of the realm of infinity itself. This is not to deny the existence of "useful" applications of mathematics. Mathematics *is* useful—it *does* have wonderful applications—but that is not why we study it. In other words, mathematics and mathematicians proceed under the principle of *contra utilitatem solam*, or "against utility alone." Again, this is not to deny the existence of marvelous applications, but it is to ensure that we are keeping "first things first." I borrow this idea from C. S. Lewis, who wrote a delightful essay titled "First and Second Things." His thesis was both simple and important. He posited that when first things are kept first and second things are kept second, both first and second things tend to work out. Conversely, if second things are placed first, it is not just the demoted first things that suffer, but even those second things that were "promoted" end up not working out. This principle applies to mathematical theory (the "first thing") and mathematical application (the "second thing"). More often than not, developments in mathematics are years,

1 This marvelous sentence was composed by my good friend Michael Austin. I have used it more times than I can count, but perhaps not yet an infinite number of times.

decades, even centuries ahead of the applications that follow. If pure mathematical curiosity is kept first, then both mathematical theory *and* mathematical application will thrive. If, however, application is held as the end-all and be-all of mathematics, then both mathematics *and* mathematical application will wither on the vine.[2]

If mathematics is fundamentally *good*, then it means that mathematics deserves to be learned precisely *as* mathematics, without necessary recourse to applications, and still less concern for "relevance" to the student. It also means that mathematical objects themselves deserve to be studied *for what they are*. For example, prime numbers are *perfectly interesting* without recourse to data encryption. Why are there so many of these basic building blocks of multiplication? After all, the basic building block of addition is simply the number 1 because any whole number can be expressed as the sum of 1s. Why is multiplication so much more complicated? Why do its building blocks (the primes) seem to occur unpredictably? Take also the simple triangle, which is perfectly marvelous on its own, independent of any structural rigidity it provides in engineering. Why is it that three side lengths form a unique triangle (assuming they form a triangle at all, which is an altogether interesting question on its own)? This is not the case with four-sided figures. With the side lengths 4, 5, 6, and 7, you will be able to make an infinite number of four-sided shapes, but take just the lengths 4, 5, and 6 and *there is only one triangle that can be made*.

It is in part because of the inherent *goodness* of mathematics that we can choose to *wonder* about mathematics. (I say "in part" because the apprehension of beauty has a lot to do with the decision to wonder.) This is because wonder is an act of the will, and the proper object of the will is the good. Granted, sometimes wonder seem to happen to us, as in those "Grand Canyon moments" in which we are overcome with awe. These are incredible moments, but this is not the normal way that wonder appears. It normally happens by choosing to *wonder about* something, by striving to *encounter* its essence. This means that, in order to find mathematics

2 I readily admit that some mathematical advances occurred because of the desire to investigate and understand a physical phenomenon, such as advancements in conics that resulted from studying the movement of the heavenly bodies. This does not contradict my thesis because it is not placing application prior to theory but rather seeking an understanding of the physical world in mathematical language, which admits a consistency between mathematical ideas and the very fabric of the universe. When mathematical properties are discovered, even if motivated by a physical phenomenon, they take on a life of their own and become worthy of study in and of themselves. This is precisely what happened with conics.

interesting, *we must choose to find mathematics interesting*. We do this by asking questions like a mathematician.

For example, I was once working with a group of second-grade teachers. They asked me to talk to them about how to make mathematics interesting for their students. I began by pointing out all sorts of compelling things about shapes and numbers. They asked, rightfully, "But how do we take our *everyday* content and make it compelling?" I then asked them what their students were studying that week, and the teachers informed me that they were working on the standard "stacked subtraction algorithm." At that point, we put a problem on the board.

$$\begin{array}{r} 492 \\ -248 \\ \hline \end{array}$$

I asked them to lead me through the process of performing the subtraction. "First," they said, "we borrow a 1 from the 9."

"Why do we call it borrowing?" I asked. "We don't give it back, right? I mean, it is more like stealing than borrowing, sort of like asking to borrow a tissue."

They laughed.

"Continue," I said.

"Well, you ~~borrow~~ steal a 1 from the 9."

"Actually," I said, "we don't steal a 1. We steal a 10, right?"

They agreed.

I asked, "Why do we do this?"

"Because you can't take an 8 from a 2."

"Really? [This is not true, but we will come back to that in a minute.] Why not?"

"Because 8 is greater than 2."

"Okay," I said, "I am satisfied. [I wasn't.] Continue."

"You steal a 10 from the 9, so the 9 becomes an 8, and the 2 becomes a 12."

$$\begin{array}{r} {\scriptstyle 8\ 12} \\ 4\not{9}\not{2} \\ -248 \\ \hline \end{array}$$

"Actually, it is a 90 that becomes an 80 rather than a 9 becoming an 8, but I digress. Please continue."

"Twelve ones minus 8 ones is 4 ones. Eight tens minus 4 tens is 4 tens. And 4 hundreds minus 2 hundreds is 2 hundreds. The answer is 244."

$$\begin{array}{r} {\scriptstyle 8\ 12} \\ 4\not{9}\not{2} \\ -248 \\ \hline 244 \end{array}$$

I thought for a minute. "Can you remind me why I had to steal the 10 in the first place?"

"Because you cannot take 8 from 2."

"Why not?"

"Because 8 is greater than 2."

"How much greater than 2?"

"It is 6 greater than 2."

"So, I only needed 6?" I asked.

"Yes."

"Why," I asked, "did I steal 10?"

"Huh?"

"Well, I only needed 6 to solve the problem, but I stole 10. So not only am I a thief," I joked, "but I am a greedy thief."

They chuckled. Nervously.

I led them through the problem again. Let's "borrow" (steal) a 6 from the 9, which really represents 90.

$$\begin{array}{r} {\scriptstyle 8} \\ 49\not{2} \\ -248 \\ \hline \end{array}$$

In the "real way" we crossed out the 9 and made it an 8. Because the 9 represents a 90, we were stealing away 10. That gave us 80, and 80 was represented by an 8 in the tens column. This time, the bookkeeping is a bit more complicated. When we take 6 away from the 90, we are left with 84. It might look something like this:

84 8
492
−248

The "8 ones minus 8 ones" is "0 ones." That is the easy part.

84 8
492
−248
0

The "84 minus 4 tens" part requires a bit more thought, mostly because the 4, being in the tens column, represents 40. Therefore, 84 minus 40 is 44. (I will squeeze the second 4 into the ones place with the 0.)

84 8
492
−248
44

Finally, the 4 hundreds minus 2 hundreds is still 2 hundreds.

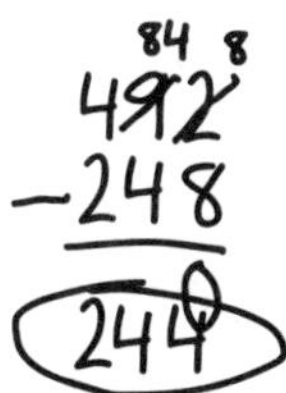

The 0 has no effect on the answer, so our difference is still 244.

"Whoa!" the three teachers exclaimed.[3]

"Wait," I said. "Can you remind me once again why I could not take the 8 away from the 2?"

"Because 8 is greater than 2."

"Yes, but I distinctly remember from algebra that I *can* take away bigger

3 It is entirely possible that this reaction is exaggerated. My own enthusiasm for this problem has contributed to my memory of the events in question.

numbers from smaller numbers. It is just that the result is a negative number. So, what is '2 ones minus 8 ones'?"

Delightfully, they wasted no time in answering, "Negative 6 ones!"

"Exactly!" And we finished the problem.

$$\begin{array}{r} 492 \\ -248 \\ \hline 25(-6) \end{array}$$

(I am wrapping the last digit in parentheses so that we do not mistake it for a subtraction problem.)

The answer is clearly "two hundred fifty-negative-six." Yes, of course, but *what is that*? It seems more complicated than Bilbo's "eleventy-first birthday." Well, "negative six ones" means we should go *down* from 50 by 6, giving us 44. So "fifty-negative-six" is really the same as forty-four. Together with the 200, we have 244. Still.

"Whoa!" they exclaimed.[4]

Then one of the teachers asked, "Do you think we should show this to our second graders?"

"No! (Well, maybe if they are a particularly curious bunch.) They are just getting their heads around the standard algorithm! This might confuse them!" I said.

At that point the teacher asked something I will never forget.

"Then why are we doing this?"

My heart sank. I paused and offered the only answer I could. "You asked me to come talk about how to instill the love and wonder of mathematics in your students. Quite simply, *you cannot give what you do not have*."

The questions about modifications to the algorithm are *precisely* the questions a mathematician would ask. This is not to suggest something is deficient about the standard algorithm. In fact, it actually helps us to appreciate its elegance. After all, as witty as the "steal only 6" idea might be, the bookkeeping was a bit difficult. The more standard strategy of regrouping a 10 is far more elegant.

Mathematics deserves to be wondered about, and that is precisely because it is *good*. In fact, you can only understand the inherent goodness

4 Ditto.

of mathematics insofar as you choose to stand in awe of it and to wonder about it. And that means pausing to ask not first about its usefulness or its relevance but instead about the *thing itself*, whether that thing is a number, a shape, or even a well-known algorithm. Do you wonder? If not, remember that it is a choice, an act of the will. There is something unique in the human soul that can only be satisfied by wondering about mathematics.

Mathematics Is True

If the good is the object of the will, which is why the goodness of mathematics is tied to our will to wonder about it, then truth is the object of the intellect. I have often thought that while all of the disciplines disclose truth, goodness, and beauty, they all do so in something of a different order and with different emphases. For example, it seems clear that the visual arts first display beauty, and through beauty they disclose truth and goodness. I think that literature first discloses goodness. (At least *good* literature discloses goodness, even if it is accomplished by disclosing the nature of good's absence: evil.) It seems obvious, then, that mathematics first presents as truth. In fact, mathematics has a unique role in our lives in the way in which it brings us to *objective* truth. Mathematics teaches us that there is objective truth in the universe and that this objective truth is *knowable*. Moreover, because mathematics is woven into the fabric of the universe it also teaches us that the universe itself is both *ordered* and *knowable*.

Objectivity in some of the other disciplines can be hard to see, but when you learn and internalize the proof of the Pythagorean Theorem, you are discovering something that is inarguably *true*. What's more, when you prove for yourself the Pythagorean Theorem, you are proving something that was true before you were born and will be true long after you are gone, and upon death it can be written on your tombstone, "This person *proved* the Pythagorean Theorem."[5] The truth of the Pythagorean Theorem is not only objective, meaning not a matter of opinion, but also *eternal*. Some statements can be objective but not eternal, such as "The strength of my reading glasses is 1.25." I can almost assure you that while this is objectively true, it is not an eternal fact, as I can already feel my vision fading with age. The truths of math, however, not only are true *now* but always will be true.

5 This is another wonderfully poetic statement for which I am grateful to Michael Austin.

There is something else, perhaps a more subtle point, about the truth that mathematics offers. Mathematics deals in the realm of ideals. While these ideals find approximations in the physical world, they themselves are not physical. Some ideals in mathematics are *numbers*. While we have seen lots of groups of "five," we have never really "seen" the thing that these groups have in common, that is, "five" in its immaterial purity. Neither have we seen numbers such as 10^{100}, which is large enough to have no possible incarnation in this universe, nor $10^{10^{100}}$, for which, even if we used a single atom to represent a zero, there would not be enough atoms to write out the number in standard form. Then there are the irrational numbers, which for different reasons we also have trouble writing out. Other mathematical ideals are shapes, such as a common circle. Mathematicians deal with perfect circles even though no such perfect circle made of matter could ever exist. What's more, mathematicians can even *say true things* about these shapes that they have neither seen nor ever will see. Finally, and perhaps most amazingly of all, mathematics deals with *infinite* things, both the infinitely large and the infinitely small. Mathematicians have no trouble adding up an infinite number of infinitely small things, and they have no trouble comparing sizes of infinity (because it turns out that some infinities are *much* larger than other infinities).

These ideals, while the property of the heavens, also find particular incarnations on earth, which is why mathematics will always find marvelous applications. These incarnations might be approximations of the ideals, but they are "real" in their own way. While the perfect circle cannot be made of matter, real material circles in their imperfect state do in fact exist. The reality is more profound: *the entire material world seems to speak of the ideal objects of mathematics*. Another way of saying this is that the material world seems to have been spoken into being using the language of mathematics. In this way, mathematics has the privilege of being a discipline that bridges heaven and earth.

While mathematics teaches us that truth itself is knowable, it also teaches us that truth is never *exhaustible*. Mathematical truth, like all truth, is a *mystery*. There are two ways to misunderstand the concept of a mystery. The first is to conceive of it as something that cannot be understood but that must be accepted by faith alone. This is to ignore the *rationality* of mystery. The second, and opposite, way is to conceive of a mystery as something akin to a mystery novel—a puzzle that can be solved once and for all, something that can be exhausted in what there is to know. Certainly, mathematics has these kinds of sub-mysteries: interesting and beautiful problems that can solved, even conquered! *Mathematical truth*, however, as a whole *cannot* be conquered. It can be known in this corner or that corner, down this hallway and up those stairs, but it is an infinite mansion with dazzling rooms behind an endless number of doors. The old cliché is true: the more you know, the more you know you don't know.

Because of the infinite mansion of mathematical ideas, it should come as no surprise that there are unsolved problems in mathematics, some of which have baffled the best mathematicians throughout history. What's more, some of the oldest unsolved problems are relatively easy to understand, and yet their solutions remain undiscovered. Since mathematics so clearly communicates objective and eternal truths, we can come to think that its work is "finished." Unlike science—which although it also discloses truth does so in a way that is "self-corrective," with new scientific advances refining our understanding using older models—mathematics rarely self-corrects. It only does so when there is an actual *error*. This permanence of mathematical truth does not mean, however, that there is not more to learn.

All told, the truth offered by mathematics is *objective*, *eternal*, *immaterial*, and *inexhaustible*. These four aspects offer the most fascinating of epic journeys to the mathematician.

Mathematics Is Beautiful

For the mathematician, the beauty of mathematics does not require a rational defense. For the non-mathematician, it can only be shown. The best proof for the beauty of mathematics is simple and self-evident.

> There is the number π. *Therefore, mathematics is beautiful.*

There are countless other proofs like this one.

> There is the Fundamental Theorem of Arithmetic.
> *Therefore, mathematics is beautiful.*

> There are the Platonic solids.
> *Therefore, mathematics is beautiful.*

The Hungarian mathematician Paul Erdős (1913–1996) said something similar.

> *Why are numbers beautiful? It's like asking why Beethoven's Ninth Symphony is beautiful. If you don't see why, someone can't tell you. I know numbers are beautiful. If they aren't beautiful, nothing is.*[6]

In other words, you either get it or you don't; there is little more to say. Of course, this does not mean that your palate for mathematical beauty cannot be trained. Just as your palate for wine can mature to leave behind white Zinfandel for the heavenly reality that is a premier cru Meursault, so too can your appreciation of mathematical beauty be developed.

What's more, there are standards of beauty in mathematics. Beauty, after all, is as objective as truth and goodness. In mathematics, it is an injustice to claim that "all correct solutions are equal." This is simply not true. Some correct solutions are better than other correct solutions because they are more elegant, more generalizable, or in many cases even *simpler*. For example, there are many ways we might learn to divide numbers, but there is something incredibly elegant about the standard long division algorithm. It is a marvel of place value and efficient bookkeeping, and it is generalizable to the long division of polynomials in high school algebra.

The poet Edna St. Vincent Millay recognized the beauty in mathematics, and she recognized it specifically in its ability to reach for the heavens, bring down eternal realities, and leave an imprint on earthly sands. She wrote the poem with which I end this chapter specifically about Euclid, but there is little doubt that her words also apply to mathematics generally.

6 Devlin, Keith. *The Math Gene: How Mathematical Thinking Evolved and Why Numbers Are Like Gossip*. Basic Books, 2000, p. 150.

EUCLID ALONE *has looked on Beauty bare.*
Let all who prate of Beauty hold their peace,
And lay them prone upon the earth and cease
To ponder on themselves, the while they stare
At nothing, intricately drawn nowhere
In shapes of shifting lineage; let geese
Gabble and hiss, but heroes seek release
From dusty bondage into luminous air.
O blinding hour, O holy, terrible day,
When first the shaft into his vision shone
Of light anatomized! Euclid alone
Has looked on Beauty bare. Fortunate they
Who, though once only and then but far away,
Have heard her massive sandal set on stone.

—EDNA ST. VINCENT MILLAY

CHAPTER 2
WHAT IS MATHEMATICS?

Mathematics is a liberal art, so a description of the nature of mathematics must begin with a discussion about the liberal arts themselves. The classical liberal arts are seven in number, but they are divided into three and four. The three are called the trivium: grammar, logic, and rhetoric, otherwise known as the "arts of language." Grammar is the art of the construction of language, and learning it allows us to piece together a coherent sentence (among other things). Logic is the use of language to piece together a proper line of reasoning. It is the *form* given to grammar. Rhetoric is the art of language that in its essence is the communication of the truth.

The word "rhetoric" needs to be redeemed. We often use it in a pejorative manner to imply "mere rhetoric" or "empty rhetoric." As an art of language, though, rhetoric is a *good thing*. It is, as Quintilian said, "the art of speaking well," or perhaps even more poignantly, "a good man speaking well" (also Quintilian). In fact, the use of language to convince others of falsehoods is not rhetoric, and not even "mere rhetoric," but rather *anti-rhetoric*. It is also significant that rhetoric completes the movement begun by grammar and logic because it tells us something about truth itself. All truth finds its proper end when communicated with others, which is another way to say that truth is both *effusive* and *communal*—it demands to be shared.

It is also significant that the arts of language are three in number. Three has always represented the eternal or the heavenly. As such, the arts of language extend in two directions. First, they extend in a vertical manner, meaning that the use of language and argumentation does not change or develop over time. A logical argument properly made will always be a valid logical argument.[7] Second, and perhaps more subtly, they extend in a horizontal manner, meaning that they are employed in all of the disciplines, from mathematics to history, from literature to philosophy, from science to economics. All disciplines use grammar, logic, and rhetoric to discover and communicate truth, which is why all teachers are professional orators.

Sitting alongside the trivium are the other four liberal arts: the

7 It is easy to convince people that the laws of logic are unchanging. It is harder to convince them that rhetoric is unchanging. This is not to suggest that rhetorical arguments do not vary depending on the audience. But the art of rhetoric itself, understood as "selecting the most persuasive argument for a given situation," together with its constructs and techniques, does not change or develop.

quadrivium. They are arithmetic, geometry, astronomy, and music. If the number three represents the eternal and the heavenly, the number four often represents the things of the earth. Together, three and four make seven, which is a number that represents "everything." At first glance, it should strike us as significant that mathematics (arithmetic and geometry) makes up half of the "things of the earth."

If we listen to the Pythagoreans, however, the plot thickens. They understood mathematics as divided into two major categories: quantity and magnitude. "Quantity" means those things that can be counted, i.e., the whole numbers: 1, 2, 3, ... "Magnitude" means those things that are made of lines, circles, and other shapes. Moreover, the study of geometry is the study of magnitude *at rest*, whereas astronomy is the study of magnitude *in motion*. This makes sense if we think about astronomy as the motion of the heavenly bodies. For our purposes, the important thing is that astronomy was really a branch of mathematics for the Pythagoreans.

On the other side of the quadrivium, arithmetic is the study of quantity *on its own*. Fascinatingly, the Pythagoreans saw music as the study of quantity *in relationship*. For those with a background in music, this makes sense. Musical notes, prior to modern standardization of frequency, were always seen in relationship to one another, and those relationships could be described with ratios. For example, if one string is half the length of another string—equivalently, if the ratio of the two lengths is 1:2—we get the musical interval of an octave. Similarly, two strings with lengths in a 2:3 ratio gives what is known as a "perfect fifth." Legend says that Pythagoras discovered this fact not with strings but rather when listening to four blacksmiths strike hammers against anvils and noticing that certain combinations sounded pleasing together. He was then able to translate these combinations into simple ratios that correspond to well-known musical intervals, now termed Pythagorean intervals. What is important here is that music, along with the rest of the quadrivium, was also seen as a branch of mathematics. Thus, I was not quite right in saying that half of the "things of the earth" are about mathematics. Rather, *all* of things of the earth are about mathematics. The ancients understood that there are two fundamental things that describe reality, language and mathematics, and that the two objects within mathematics are number and shape.

MATHEMATICS			
Quantity (*The Discrete*)		Magnitude (*The Continuous*)	
Alone (*The Absolute*) Arithmetic	In Relation (*The Relative*) Music	At Rest (*The Stable*) Geometry	In Motion (*The Moving*) Astronomy

This schema gives us a sense of how to situate mathematics within the context of the classic seven liberal arts, but it only begins to answer the question, "What is mathematics?" The Pythagorean division gives us a sense of the *content* of mathematics (number and shape), but it does not answer the question, "What is the mathematical act—what does it mean to *do* mathematics?" In other words, it is one thing to know that mathematics is the study of quantity and magnitude, but what does it mean to *study* this content?

I have a theory that has been field-tested on occasion, much to the chagrin of my children, who are usually present during the field-testing. The theory is that if I ask random people on the street what a biologist does, most of them will be able to offer a fairly accurate description. While they may not understand advanced biological concepts and may not know all of the details of a biologist's work, most people have a decent mental image of what it means to *do* biology, even in its various subdisciplines, e.g., cellular biology, field biology, etc. The same thing is true of a variety of careers: history, writing, etc. When I ask people to describe the life of a mathematician, however, I rarely get an answer that captures the essence of this career. I usually get one of three incorrect answers. The first is the image of someone who spends his days figuring out creative ways to multiply large numbers together. We might chuckle about this, but it turns out that it is closer to being correct than the other two answers. It at least has the element of serious play with actual mathematical objects that we saw in our non-standard subtraction algorithms. A second answer is a description that more accurately describes an engineer: using math to build structures and solve physical problems. The third answer describes a statistician: someone who collects and analyzes data. Both of these are noble careers, but neither offers an accurate description of a mathematician.

How is it that most people have had thirteen years of mathematics classes from kindergarten through the twelfth grade, and more if they took college classes, and yet have trouble describing what a mathematician does? Of course, I am taking for granted that a mathematician does mathematics. The only possible answer is that in those thirteen years of mathematics classes, most people have not actually done much mathematics. This is not to suggest that they have not studied mathematical topics. As we saw, the *content* of mathematics consists fundamentally of numbers and shapes, and these are certainly covered in elementary, middle, and high school courses. But simply learning facts and skills within a discipline is not the same as learning the *discipline itself*, and still less is it *doing the discipline*. Think about learning the names of notes, names of intervals, and even advanced techniques of chord progressions but never experiencing a finished piece of music and still less producing a piece of music either on paper or through the use of an instrument. It could hardly be said that one has *learned music*.[8] Yet this is precisely the case for most people who have survived mathematics curricula over the course of their time as a student. What I posit is that we were all taught *computation* and *application*. We were taught to perform operations on numbers, functions, and perhaps even matrices; we were taught to measure certain things about shapes, such as perimeter, area, and volume; and we were taught to apply some of these ideas to "real-world" phenomena.

I will resist the urge to go on a full rant against the push for "real-world problems" in the mathematics textbook industry. I will instead offer three thoughts about applications (word problems). First, learning mathematics is fundamentally about *learning mathematics*, not about learning to apply it. There is nothing wrong with a word problem, especially for young students, but we must understand its proper place and purpose. When learning what it means to add numbers together, a young student will find it helpful to think about "combining three apples with two apples" and wonder about how many apples there are in total. It is better still if young students have actual, physical objects on their desks to hold and move and use to model addition. Even to young students, however, we

8 Paul Lockhart, in *A Mathematician's Lament*, presents an extended description of this analogy.

do not teach the word problems to show the usefulness of mathematics, as if someday a child might be standing with two apples in one hand and three in another and say, "I wonder what operation I can use to figure out how many apples I have." The truth is, the child will probably just count the five apples—or better yet, *recognize* the shape of the number 5—and not worry about the choice of operation, still less about the actual computation. The real value of the word problem is that it allows students to contextualize the abstract concept with which they are wrestling, and it is these abstract concepts that form the real stuff of mathematics. We want the students to understand *addition itself*, not the application about apples, and the word problem helps young students move from something familiar to the abstract. In other words, mathematics is not at the service of the word problem; instead, the word problem is at the service of mathematics.

Second, these "real-world problems" are almost never about the real world. Suppose two cars start from the same point and head in two different directions, one north and one east. If one car travels at 45 miles per hour and the other car travels at 60 miles per hour, how far apart will they be after six hours? How "real" is this problem? How often have you driven a car at *exactly* 45 miles per hour for a sustained six hours? How often have you and a friend traveled in two directions that form an *exact* right angle and maintained that direction the entire trip? How often have you started off with a friend from *exactly the same place*? The cars would have to be on top of each other! The term "real world" is not quite fitting for this situation. Again, this does not mean students shouldn't work on these sorts of idealized problems; it simply means that the purpose of a problem like this is not to show the application of mathematical concepts. Rather, the purpose is to contextualize the abstract. In this case, the word problem is teaching something about how traveling at different speeds in different directions affects the position of two objects, and it reinforces the importance of the Pythagorean Theorem. Later, in calculus, it will say something about change itself and how related quantities change with respect to one another. The marvel of the fact that mathematics *can* be contextualized lies in its ability to bridge heaven and earth.[9]

9 In fact, the beauty of the word problem is found precisely in its idealized nature, not in its approximation within the material universe, which in some ways makes it more real than its material counterpart. There is a parallel here to literature, especially children's literature. The hero and the villain, those examples of pure good and evil, are in some ways more real than more "realistic" characters who have, as Aleksandr Solzhenitsyn said, the line of good and evil running through each of their hearts.

In other words, the beauty of the problem has more to do with its being a "real problem" than a "real-world problem."

Third, if mathematics is not first about real-world problems, and if these word problems are not actually "real," why is there such a push in the textbook industry to include them as upward of 60 percent of the content? The usual answer is that we need to make mathematics "relevant" to students. Even given the idealized nature of these problems, the claim is that they demonstrate the usefulness of mathematics for students and thereby motivate them to learn it. We only need to ask ourselves, after thirteen years of mathematics classes that are designed to make mathematics relevant, how many people come out of high school motivated by mathematics, let alone deeply in love with it? The reason for this failure is simple. *Most of us will never use most of the content we learned in our mathematics classes.* Apart from professional mathematicians, mathematics teachers, and perhaps those in a handful of other careers, most people will never even use something as important as the Pythagorean Theorem. This is all to say that the utilitarian push for mathematical application in the pursuit of relevance serves only to present students with mostly irrelevant examples. It fails to do the very thing it claims to do. What *does* work, what *does* inspire students, what *does* cause them to fall in love is to help them to encounter mathematics in all of its purity and beauty. This is not different from other disciplines. The content of mathematics is perfectly interesting on its own, as are all things true, good, and beautiful. We need only have the eyes to see it and a teacher who can unveil it. It is also not different from falling in love with a person; we do not perform a utilitarian calculation about the other's relevance to our lives. Instead, we must have an authentic encounter.

Enough about what the art of mathematics is *not*. Let us turn our attention to what it *is*. First and foremost, mathematics is the study of real things and their properties. Mathematicians look at objects, be they numbers, shapes, or other more abstract structures, and they look for patterns in these structures. The patterns lead to questions about the universal properties of these structures, which in turn allow the mathematician to pose problems. In solving the problem, the mathematician is producing a logical argument that comes to fruition in the *proof*. We could therefore say that the art of mathematics is an art of *properties*, *patterns*, *problems*, and *proofs*. The proof is the end of the mathematical process. It is the formal act of *mathematical rhetoric*. Remember that rhetoric is the art of bringing others to truth: it is what prevents the first two arts of language, grammar and logic, from becoming activities merely for the individual. Rhetoric demonstrates that language is an essentially *communal* activity. It recognizes that truth, upon discovery, demands to be shared. The mathematician shares a discovery and a logical argument, demonstrating it through the mathematical proof. It is also the case that proof is the *poesis* of mathematics, the creative product of the mathematician, in a similar way that the painting is the *poesis* of the painter, the poem of the poet, and the story of the author.

The fact that mathematics ends in an act of rhetoric is why a teacher should insist that a student "show his work." It is not so that the teacher can "see the thinking" of the student, and still less is it so that the student can "check his work." Rather, it is because the student has discovered something to be true, and he should now have a desire welling up from within to share this creation with others. If a student solves a problem mentally, the teacher's response should not simply be "You must show your work." Instead, the first response should be "Well done, my good and faithful mathematician. Now convince others of the truth you have discovered and put it down in writing for all of posterity!"

For the mathematician, the case is no different. He is a professional "proof writer." This is why Euclid's *Elements* is a marvelous text to use for high school geometry: it is a formal, completed act of mathematics that happens to be accessible at that age. As an original source, it also demonstrates for students that mathematics is a human endeavor, embarked upon by real men and women. I am in good company in recognizing the unique importance of proof, and specifically that found in Euclid, for one's education. As a young law student and lawyer, Abraham Lincoln carried a copy of Euclid in his satchel. In 1864, the *New York Times* printed a conversation between Lincoln and the Reverend J. P. Gulliver.[10] Lincoln remarked on his own education in the law: "In the course of my law-reading I constantly came upon the word demonstrate. I thought, at first, that I understood its meaning, but soon became satisfied that I did not." After perusing several sources and remaining unsatisfied, Lincoln said

10 Gulliver, J. P. "*Mr. Lincoln's Early Life: How He Educated Himself.*" New York Times, September 4, 1864. Retrieved from https://www.nytimes.com/1864/09/04/archives/mr-lincolns-early-life-how-he-educated-himself.html on January 18, 2020.

to himself, "You can never make a lawyer if you do not understand what demonstrate means." Once he realized this, he said, "I left my situation in Springfield, went home to my father's house, and staid there till I could give any propositions in the six books of Euclid at sight. I then found out what 'demonstrate' means, and went back to my law studies."[11]

Gulliver echoed Lincoln's praise for the *Elements*:

> *No man can talk well unless he is able first of all to define to himself what he is talking about. Euclid, well studied, would free the world of half its calamities, by banishing half the nonsense which now deludes and curses it. I have often thought that Euclid would be one of the best books to put on the catalogue of the Tract Society, if they could only get people to read it. It would be a means of grace.*[12]

Lincoln responded, laughing, "I think so. I vote for Euclid."

In the 1858 Lincoln–Douglas debates, Senator Douglas issued an *ad hominem* attack against Senator Lyman Trumbull. Mr. Lincoln responded to the attack by referencing the Geometer:

> *If you have ever studied geometry, you remember that by a course of reasoning, Euclid proves that all the angles in a triangle are equal to two right angles. Euclid has shown you how to work it out. Now, if you undertake to disprove that proposition, and to show that it is erroneous, would you prove it to be false by calling Euclid a liar?*[13]

Those who have not studied Euclid or delved into other examples of pure mathematics may ask, "What is a proof?" I have categorized it as a formal act of rhetoric, and Lincoln describes it as a demonstration, but I fear that our own experience of proofs in high school geometry has left many of us with a sour taste in our mouths for them. The first reason for this is that the proofs in most geometry textbooks deal with situations so specific that one wonders why we should care about them at all. For example, in the following diagram, A is the midpoint of $\overline{BC}$ and $\overline{DE}$. Prove that $\overline{BD}$ and $\overline{EC}$ are parallel.

---

11 Ibid.

12 Ibid.

13 Angle, P. M. *The Complete Lincoln-Douglas Debates of 1858*. Chicago, IL: University of Chicago Press, 1991, p. 275.

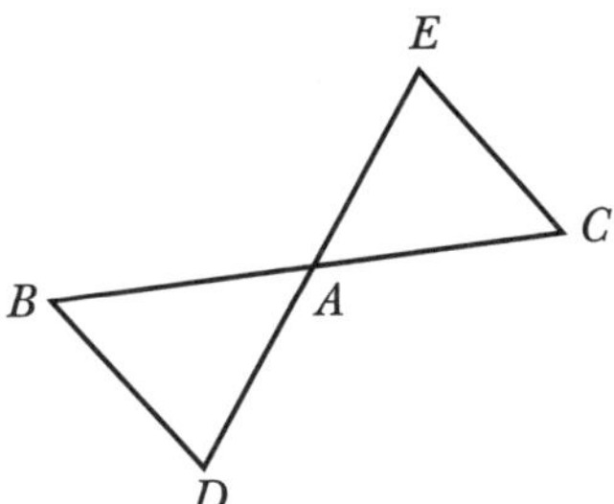

In reading this problem and looking at the diagram, we wonder why we should care. It is a very particular drawing of a very particular situation. Mathematically, it is a bit dull and fails to draw us in. Now, to be sure, there *are* interesting things about this problem about which we can wonder. Remember, wonder is an act of the will. For example, "Why is it that whenever two line segments meet at their midpoint, if we join the 'top' and 'bottom' of the X that is formed, these two new lines are always parallel?" Stated this way, the problem is about an interesting property of intersecting lines. It is a commentary on real and universal things.

Authentic mathematics is the discovery of interesting and universal properties of the objects under study. The subsequent proof confirms these properties. Actually, the proof does more than that. It *elicits and communicates understanding* of the property, or at least a *good* proof does. Consider, as an example, a remarkable property of odd numbers. Let's start by making a list of the sum of consecutive odd numbers beginning at 1.

$$
\begin{aligned}
1 &= 1\\
1+3 &= 4\\
1+3+5 &= 9\\
1+3+5+7 &= 16\\
1+3+5+7+9 &= 25\\
1+3+5+7+9+11 &= 36
\end{aligned}
$$

Do you notice something about the sums? Each is a perfect square. (A perfect square is the result of a whole number multiplied by itself, e.g., $36=6\times6$, $25=5\times5$, $16=4\times4$, etc.) In fact, the sums represent *all* of the perfect squares, in order, starting at 1. The natural questions are "Will this pattern continue indefinitely?" and, if so, "Why?"

A proof, by nature, has to be logically correct: it needs to demonstrate, step by step, the thing being claimed. But that requirement is only a bare minimum. As we said earlier in our discussion of beauty, some solutions, or proofs, are better than others. A beautiful proof will elicit and communicate understanding. Paul Erdős, whose comparison of the beauty of numbers to that of Beethoven's Ninth Symphony I quoted earlier, knew this well. One of the most eccentric mathematicians of his age—which is saying something!—he spent most of his adult life living out of a suitcase. He would show up at departments of mathematics in universities around the world and exclaim, "My brain is open!" Mathematicians would sit down and discuss their work, and he would offer ideas and pose interesting questions. He thrived on working with other mathematicians and coauthored well over one thousand papers. Thus he became one of history's greatest mathematical collaborators. In fact, it has become a source of pride for mathematicians to note how closely they have published with Erdős. A mathematician can refer to his or her "Erdős Number." If you published directly with Erdős, then your Erdős Number is 1. If you published with someone who published with Erdős, then your Erdős Number is 2. (This is something of a mathematical "Kevin Bacon" game.)

Erdős loved elegant proofs. Often a theorem would be proven, or accepted as proven by the mathematical community, and yet Erdős would say, "Yes, but that is not the proof from *The Book*." His idea was that while mathematical theorems had numerous correct proofs, only one was in *The Book* held by God. The proofs in *The Book* were so elegant that, upon reading them, one fully understood not just *that* the theorem was true, but *why* the theorem was true.

Returning to our observation about the sum of consecutive odd numbers, there are any number of ways of showing that this is true. A common way would be to employ an advanced technique such as the Principle of Mathematical Induction. While this is a valid proof, my guess is that if I were to present it here, some readers would struggle to understand it, and most readers would fail to see *why* the property holds, even if they were convinced *that* it holds.

A better proof—whether or not it is from *The Book* we leave to the judgment of Erdős—would be to start with a single dot, representing the "1" in the sums.

We can add three dots onto this in any number of ways, but I suggest that we "surround" the existing dot on two sides.

Where shall we put the next odd number, 5? Again, I suggest that we surround the existing four dots on the top and the right.

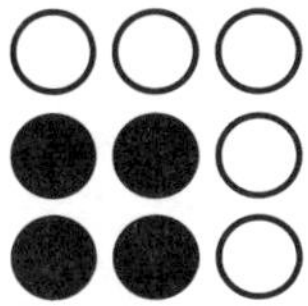

We can see that the total is 9 ($1+3+5=9$), but we can also see that the shape is a *square*, in fact a three-by-three square. Geometrically, it is clear now that $1+3+5=3^2$. This is also a good reminder that the numerical operation of "squaring" is called that precisely because it is based on the geometry of a square! We can then continue this pattern. Wrapping the next odd number, 7, around the top and right side of the three-by-three square would complete a four-by-four square. The geometrical argument here does a very good job of showing *why* the sum of consecutive odd numbers starting at 1 always generates a perfect square. The dots can be arranged in a way that *forms a square* because the next odd number is simply the next layer in the picture. This narrative, together with the instructive pictures, is a *mathematical proof*, and a beautiful one at that.

As a sidenote, it is worth pointing out that this proof was not written in a two-column format. While a decent pedagogical tool for starting students off in their proof-writing journey, the two-column proof often does little to disclose the *why* of a proof, and still less the beauty of the result. Using a two-column proof is like solving a corn maze by knowing the starting point, the ending point, and the rules, such as that one cannot pass through the corn walls. We can "start with the given," meander, make logical steps along the way, and end up, almost by luck, at the "prove" statement. In doing so, the "correct" proof discloses nothing of the truth we are trying to demonstrate. Many a reader may recognize, perhaps with a nervous twitch, this general formula for success from high school geometry: start with the "given," end with the "prove," and good luck along the way! In contrast to this wandering process, encountering a proof from *The Book* is like looking down on the corn maze from above and seeing the way out. A description of the solution is still required and turn-by-turn directions should be offered, but these become details in the main "narrative of the *why*." The narrative of *why* consecutive odd numbers sum to squares is precisely that squares themselves can be decomposed into odd numbers that form the L-shaped layers of the square.

We end this chapter by returning to mathematics as a liberal art and the pervasive question of utility. It is most often in mathematics classrooms that teachers hear the question, "When are we ever going to use this?" If we view mathematics in the right way, the answer should be as simple as it is for the question of why we study Shakespeare. It is not a utilitarian calculation. We study these things because they free the mind to think in a way that is unique. Shakespeare is unique and irreplaceable in one's education, and so are the art and truths of mathematics. This "freeing of the mind" is why the seven liberal arts are called the *liberal* arts. The word comes from the Latin word *liberare*, meaning "to set free." The *fine* arts have as their "end" a product that is outside the artist: the painter produces a painting, and the sculptor a sculpture. The *liberal* arts, on the other hand, have their end in the human mind itself. This is not to suggest that the liberal arts are not *creative*. After all, we already described proof as the *poesis* of the mathematician. But it *is* to suggest that the creative act found in the liberal arts is fundamentally about freeing the mind to *know* and to *know in a particular way*. The liberal arts are about freeing the human person to be more fully human. If I am correct in saying that mathematics satisfies something unique in the human soul that yearns to know and to know in a way that can only be found in mathematics, then it is also true that the human person is incomplete without mathematical knowledge and mathematical wondering.

CHAPTER 3
THE QUEST FOR A BETTER MATHEMATICS CURRICULUM

As I said in the introduction, this book is not a curriculum. Still less is it an extended commentary on the typical K–12 mathematics curriculum found in schools. I also wrote that the audience for this book consists not only of mathematicians or mathematics educators but also anyone who has a mere spark of curiosity about this beautiful art form. That said, the book does, by the nature of its content and purpose, offer a critique of the presentation of mathematics in most K–12 programs. This should be of some interest to a broader audience. After all, everyone has some memory, whether positive or negative, of mathematics as they learned it in elementary, middle, and high school. And some of my readers will be educators, mathematicians, curriculum designers, policymakers, or assessment creators. If you are one of these people, consider this book the beginning of a conversation.

I am often reminded that I went through an earlier iteration of the K–12 curriculum, and through it I fell in love with mathematics, or at least "in like" with it. I owe much to the scope and sequence that formed me, though I certainly owe more to the exemplary teachers of mathematics who brought it to life for me, including my own father, who was my precalculus teacher. But the true discovery of my passion for mathematics came in college, when I first experienced the discipline's real nature: *proof*. After making my way through K–12 mathematics courses, studying mathematics in undergraduate and graduate school, and teaching mathematics for a decade, I have a small number of humble suggestions to offer for improving the current mathematics curriculum in our schools. I beg the reader's patience in allowing me to outline these before getting to the "good stuff."

1. Reduce the Content

There has *always* been too much to cover, and it seems like every time there is a national effort to write a new set of mathematics standards, everyone admits that there is too much to cover. Then a committee is assembled that produces a curriculum that has more to cover than the previous iteration did. I don't say this to be cynical; I just mean to point out an unfortunate reality of "standards" work. When mathematics is reduced to isolated student objectives rather than expanded into a compelling mathematical narrative that communicates *what mathematics is*, various people inevitably will lobby for the importance of various pieces of content. The obvious result is a scope and sequence that is both fragmented and disproportionately wider than it is deep. Don't get me wrong: content *is* important because truth itself is important, as is a breadth of experience in a variety of mathematical subdisciplines. But it is also important to be able to sit with the content, wonder about it, and deeply internalize its truth and its beauty, and it is critically important to be able to use the content to engage in authentic acts of mathematics.

The much more subtle problem with mathematics standards is that of fragmentation. The current standards make it nearly impossible to construct courses that have compelling mathematical narratives. For example, one way to think about an Algebra I course is through the story of *symmetry*, a story that culminates in the quadratic formula. The strands of data analysis woven into algebra by every set of state and national standards are difficult to fit into this narrative. This is because data analysis is not really part of algebra, but rather belongs to the entirely different subdiscipline of statistics. Of course, the misfit is more egregious when some of the data analysis standards are moved into geometry. I don't mean to be hard on this particular strand; it's just one example. The reality is that the vast number of standards across several strands cause the dual problem of breadth being prioritized over depth and the difficulty of giving a course any coherent narrative meaning. While this suggestion likely will not get much support among policymakers, we would do well to cut the number of topics covered in half.

2. *Give Proof Pride of Place at All Ages*

Attention should be given to proof and argumentation at every stage, from kindergarten through calculus. This is not to suggest that students should be writing formal mathematical proofs in elementary school, but it is to say that they should be studying real mathematical objects (shapes and numbers), asking interesting questions about them, and struggling to understand *why* certain things are true. Take, for example, the very first time students encounter the area of a triangle. They probably already have the sense that area means "the number of 1×1 squares that can fit into the shape," and they have probably already noticed that we can find the areas of rectangles by thinking about them as rows and columns, so a 3×6 rectangle can be thought of as "3 groups of 6" (or "6 groups of 3") and therefore has an area of $3\times6=18$ square units.

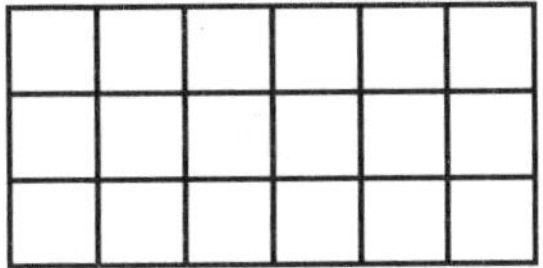

The problem with a triangle is that we have to start chopping up the squares to try to get them to fit.

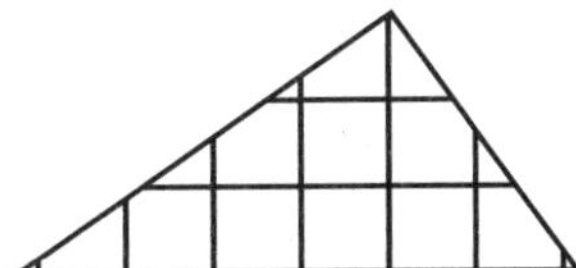

While thinking outside the box is often good advice for problem solving, sometimes it is worth thinking inside the box.

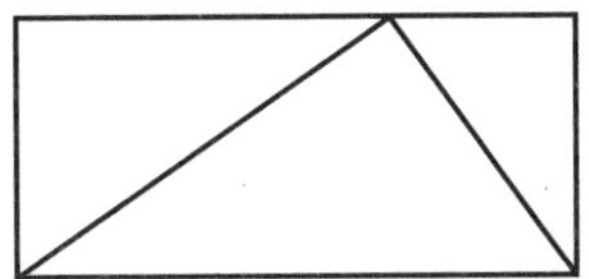

What fraction of the box is filled by the triangle?[14] To help us see this answer, we draw a line.

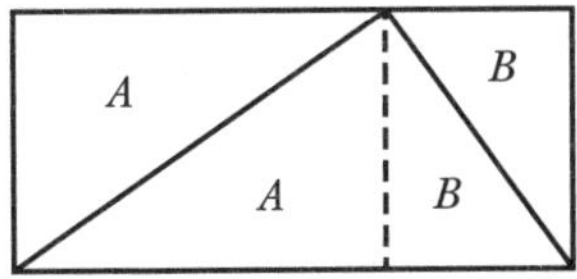

The regions marked *A* are identical because the left side of the triangle cuts the left box in half. Similarly, the regions marked *B* are identical because the right side of the triangle cuts the right box in half. Therefore, the triangle makes up exactly half of the area of the box. Because the area of the box is

14 Paul Lockhart uses this example in *A Mathematician's Lament*.

"length times width," or $l \times w$, it follows that the area of the triangle must be $\frac{1}{2} \times l \times w$. Of course, the more well-known formula for the area of a triangle is $\frac{1}{2}bh$, or "one half base times height." It is unfortunate, actually, that we refer to "length" and "width" in the rectangle formula but "base" and "height" in the triangle formula. If we had been consistent, when students see the area formula for a triangle, they would literally *see* that it is "one half of a rectangle."[15]

If students are a bit older, they can work with just the picture of the triangle and the box. Younger students can cut the diagram apart and rearrange the pieces of paper to see that the triangle really is "half of a box." Regardless of how it is done and how many details are layered in, the *idea* in this proof surely qualifies it as a "proof from *The Book*"—and it is accessible for both the young and not-so-young.[16]

3. Reduce the Amount of Algebra

There are three areas of mathematics that deserve to be covered in a mathematics curriculum: number theory, geometry, and algebra. Elementary school mathematics should prepare students for study in each of these areas, but starting in middle school the three subdisciplines should get equal attention. As it stands, geometry is regulated mostly to a single year of study—with a few topics randomly tossed into other courses—and number theory gets barely any time at all. The rest of the middle school and high school curriculum consists of algebra, whether we call it Algebra I, Algebra II, Precalculus, or something else. In fact, this emphasis on algebra is intentional and is felt as early as elementary school when "research-based" programs tout measures of "algebra readiness" as a mark of their success. We should ask ourselves, at the very least, why they do not also tout "geometry readiness" or "number theory readiness," or even (gasp) "proof readiness."

15 This proof only works when the two angles on the base are acute. I leave it to the reader to figure out how the proof might be adjusted if one of the base angles is obtuse. You might want to start with a triangle that is "slightly obtuse" before trying out your argument on a "very obtuse" triangle.

16 This is a perfect example to illustrate the possibility of increasing rigor for the same proof at a variety of ages. As stated here, perhaps without recourse to variable names, it is appropriate for elementary school students. In a high school geometry class, we would ask questions like "How do you know that the two triangles marked *A* really are identical?" Students can then fill in more details that would not be accessible at younger ages, thereby generating a more mathematically rigorous proof.

If the mathematical act is about *proof*, then geometry and number theory need to be given a much more prominent place, and proof should drive the curriculum in these two areas. Formal proof can be introduced much earlier if number theory is studied starting in middle school. Students can explore properties of divisibility, prime numbers, and other types of numbers such as "perfect" numbers, and they are fully capable of arguing the truth of these properties. What's more, the proofs in this area can be the impetus for introducing algebraic concepts. After all, if I am trying to prove that two numbers being divisible by a third number means that their sum is also divisible by that third number, it would be really helpful to have variables that can represent an infinite number of cases, e.g., "If a and b are both divisible by c, then $a + b$ is also divisible by c." As the curriculum is currently sequenced, students have little motivation to learn how to manipulate expressions and solve equations.[17]

Geometry should be given much more than one year of study. There are countless important topics deserving of attention that are not traditionally covered because of a lack of time: the Platonic solids, the construction of the regular pentagon, the geometry of the golden ratio, and more. Like number theory, geometry should be driven by proof. Students should explore the reality of shapes, work to draw conjectures, and prove their results. Remember, without proof, the process might be about exploration or even problem solving, but it is not a full act of mathematics. The appropriateness of proof at each developmental stage should drive the curricular choices. When looking for content, we should *first* ask, "What kinds of proofs are students able to write at this age?" Through this lens, it is not at all clear that eighth-grade algebra, such a precious mark of an "advanced curriculum," is the right choice for *any* student, let alone for *all* students. Number theory might be a more appropriate choice because it extends nicely from their work in elementary school and paves the way for algebra itself.

17 When I speak of the need to properly "motivate" the learning of mathematical concepts, I'm not talking about using "real-world" problems to make mathematics relevant. The pure math concepts themselves are only interesting problems when pedagogically motivated by a sense of why someone should wonder about this at all. Wonder is a choice, but it is the task of the teacher to present the topic in such a way that helps students to make that choice. Why would we start learning to factor trinomials apart from a need to do so? But if a polynomial is seen as a "base x" number, then breaking it down into factors is interesting. Are there "prime" polynomials like there are prime numbers? Do polynomials have a unique factorization into primes just like numbers do? These are interesting questions that can properly motivate us to begin investigating the factoring of polynomials.

4. Integrate Original Sources

I have already mentioned the unique value that Euclid's *Elements* can bring to the study of geometry. There are other original sources that should be used, for example works by Cauchy, Archimedes, Descartes, Cantor, and Newton. Not all of these are appropriate at all ages, of course, and judicious decisions need to be made as to when and how to employ them. Employing them, however, allows students to enter into the minds of the masters of mathematics by reading the best that has been written. It also allows them to understand that mathematics is the result of real men and women investigating real objects and writing real proofs. Mathematics comes from mathematicians, not from textbooks, and certainly not from the authors of national standards.

5. Establish a Canon of Theorems and Proofs

There are important theorems and proofs that *everyone* should learn. Mathematics is filled to the brim with remarkable and beautiful results. A mathematics program, particularly at the middle and high school levels, should focus less on skills and more on pure mathematical truth. Just as careful thought should be given to crafting a canon of literature and philosophy that all students should read, the same care should be given to thinking through the theorems and proofs that everyone deserves to experience. The theorems in this book are not by any means an exhaustive list, but they do represent some of the best of what mathematics has to offer, and more to the point, they are widely accessible. In part, this book is meant to start a conversation about a "mathematical canon," but by no means do I think that it presents a *definitive* canon.

6. Redefine Success

This brings me to the last and most important aspect of a good mathematics curriculum. I am quite serious when I say that *the mark of success in a mathematics program should be first and foremost the degree to which all students fall in love with mathematics*. We simply must move beyond the push for standards-based, mastery-driven education that has led to the so-called "data-driven instruction" movement, in which only short-term results, even if "short-term" means a full year, are measured and valued. What is wrong with exploring a deep and complicated concept for *years* if it is beautiful and worth learning? What is wrong with admitting that some things, often the most important things, cannot be mastered, let alone measured? (I think all high school students should read Aristotle's *Nicomachean Ethics*, but I am certain that none of them will master it.) It may be that data-driven instruction has the potential to increase scores on standardized assessments, but we should ask the essential question: *Are these assessments measuring the right things?* Where on the state tests, the nationally normed benchmark exams, the SAT and ACT, or really any standardized assessment do we find the ability of students to sit with a difficult problem and come out on the other side with a well-written proof? Assessments that do not do that are not assessing mathematics, and they are certainly not assessing the *love* of mathematics.

I repeat what I said at the start of this part of the book. There is something unique in the human soul that can only be satisfied by wondering about mathematics. So before delving into the best of what mathematics has to offer, ask yourself, "Do I wonder? Do I wonder about mathematics and mathematical realities? Why are certain things true? Why do things work the way they do?" Remember, wonder is an act of the will.[18] It is a choice. So, when it comes to mathematics, *do you wonder?*

18 As an act of the will, wonder is like love. It is not a passive emotion that we fall in and out of, but rather it is a choice. And wonder and love have a lot to do with one another. I love my wife insofar as I stand in great awe and wonder of her. I know she feels the same way. She says all the time, "Husband, I often wonder about you."

CHAPTER 4
WARM-UP: A MYSTERIOUS FACT ABOUT THE ANGLES IN A TRIANGLE

BEFORE JUMPING into our official list of proofs, let's warm up with a geometric fact taught as early as elementary school. We all know what a triangle is, and we likely remember something about how the three angles add to 180°. It is worth pausing, however, to note that it is not immediately obvious that the angles in all triangles will even add to the *same number*, let alone to the specific value of 180°. If one hundred people draw one hundred triangles, measure the three angles, and add up those values, why would we expect all of them to get the same total? After all, the same exercise doesn't work for the side lengths of the triangle. If those same one hundred people were to measure the three sides of their triangle and add up the lengths, they would assuredly get very different totals.

The fact that the sum of the angles is always the same is interesting; the specific value's being 180° is nothing short of astounding. To see why, consider that when Euclid, the greatest geometer ever to live, proved this result, he did not deal with degrees. In fact, his only unit of angle measurement was "one right angle." Therefore, rather than talking about the angles in a triangle adding to 180°, Euclid said, "In any triangle ... the three interior angles of the triangle are equal to two right angles." Two right angles will form a straight line, and so the theorem about the sum of the angles in a triangle is phrased accordingly.

> *Triangle Angle Sum Theorem*
> In any triangle, the three angles can be rearranged to form a straight line.

While a well-conceived proof elicits understanding, a well-phrased theorem can sometimes point the way. When we phrase the Triangle Angle Sum Theorem this way, we get a hint as to how to go about searching for a proof. Rather than trying to find a way to "add up angle measures," we are now looking for a way to rearrange the angles to form a straight line. To begin our search for a proof, we start with a triangle.

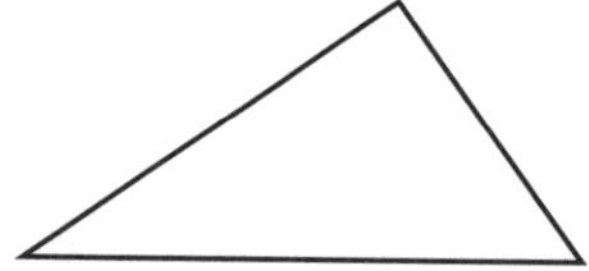

It is important to note that we are trying to prove something about *all* triangles, yet we have only drawn one *specific* triangle. One way to prove something about all things in a category is simply to test it out for *all of those things*. For example, if we were to claim that "Any two positive numbers less than 4 will always have a product that is less than 10," we could simply test this case for all possible pairs of numbers.

$1 \times 1 = 1$	$2 \times 1 = 2$	$3 \times 1 = 3$
$1 \times 2 = 2$	$2 \times 2 = 4$	$3 \times 2 = 6$
$1 \times 3 = 3$	$2 \times 3 = 6$	$3 \times 3 = 9$

In every case, the product is less than 10. (If you are familiar with the Commutative Property of Multiplication, you can cut your work almost in half because 2×3 is the same as 3×2.)

This is not a very elegant way to prove something that has a larger number of cases, but it *is* valid. Can we proceed this way for the Triangle Angle Sum Theorem? Can we draw every possible triangle, cut apart the angles, and rearrange them onto a straight line? There are at least three problems with this approach. The first problem is the impossibility of drawing a perfect triangle to begin with. No matter how straight our edge, the triangle we draw will be imperfect if only because the lines drawn have a certain thickness, and a true triangle's sides are what Euclid would call "breadthless lengths." A second problem is rearranging the angles to see that they form a perfectly straight line. How would we know if our angles actually add to 180° and not, say, 179.999936275°? The third problem is that there are an infinite number of possible triangles. We literally cannot draw them all. This raises the question, "How *does* one go about proving something for all things in a category when there are an infinite number of things in that category?" (A similar problem would exist with numbers if we were trying to prove something like "The sum of two odd numbers is always even.")

The problem of proving something for an infinite number of cases is a problem we are going to encounter over and over as we make our way through the various theorems in this book. While we are only drawing one representative triangle, we need to take great care to ensure that our argument does not depend on any particular properties of the specific triangle we are using in our diagram. For example, the triangle we have drawn seems to have angles less than 90° and has sides of three different lengths. We will want to make sure that no part of our argument depends on these facts. When we proceed carefully, it is a marvel of the mathematical process that we can provide proofs not only for objects that are "ideal" and cannot actually be drawn but also for an entire infinite collection of these ideal objects.

In other words, even though none of us has ever drawn or seen a perfect triangle, and even though there are an infinite number of these perfect triangles, we can still *know things* and *prove things* about all triangles. That, my good and faithful mathematicians, is amazing.

With that, let us attend carefully to our representative triangle.

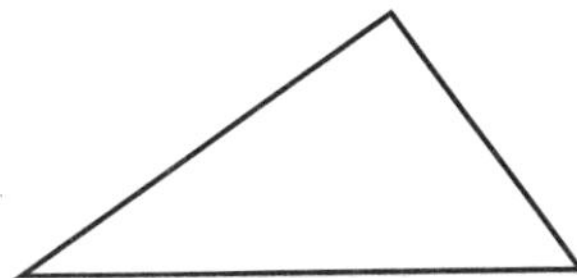

Our task is to rearrange the angles to form a straight line. We have three lines[19] with which we can work, but none of these lines has three angles that are next to each other. In the face of this dilemma, as the mathematician James Tanton often says, "If there is something in life you want, make it happen! (And deal with the consequences.)" This is not to suggest that mathematics is a matter of simply pretending that things are true, and still less is it to suggest that the truths of mathematics are somehow relative or a matter of wishful thinking. The point is that mathematics is inherently *creative*. Most of the famous and interesting proofs have profound moments of creativity, moments when the reader says, "How did you know to do *that*?"

19 Modern geometry texts differentiate between *lines*, which are infinite in length, and *line segments*, which are finite. The edges of a triangle are, using this vocabulary, *line segments*. Euclid and others do not make this distinction, however, instead opting for "straight line" and "straight line produced indefinitely."

Our creative moment is to draw a line on top of this triangle that is parallel to the bottom side.

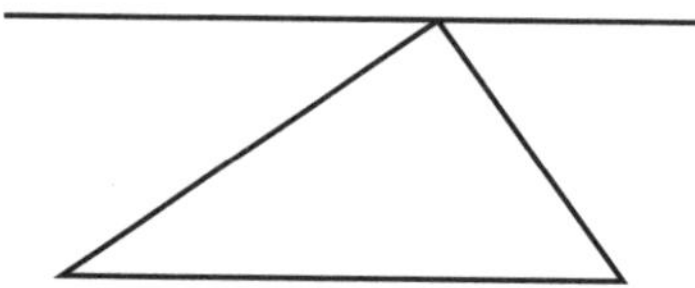

The beautiful and insightful thing about this move is that we now have three angles nicely situated on a straight line. The question is "Did we get lucky? Are these three angles identical to the angles of the triangle?" The answer is yes. The three angles along the top line are in fact identical to the three angles found in our triangle. That said, "lucky" is not quite the right word here. It is not so much luck as it is the reason *why* our theorem is true. But the feeling of luck points precisely to the creativity of the proof and the surprising result of the theorem itself.

Of course, our proof is not complete. We cannot simply state that the three angles on the line are the same as the three angles in the triangle. Such a fact might be obvious for the middle angle, which also happens to be an angle from the triangle. It is not so obvious for the other two angles. In order to argue this, we will take advantage of a *lemma*. In mathematics, a lemma is a mini-theorem that helps to prove a larger theorem.

> *The Z Lemma*
> Whenever we see lines that form a Z, with the top and bottom being parallel, the angles formed are identical.

A picture is worth, at least in this case, about twenty words.

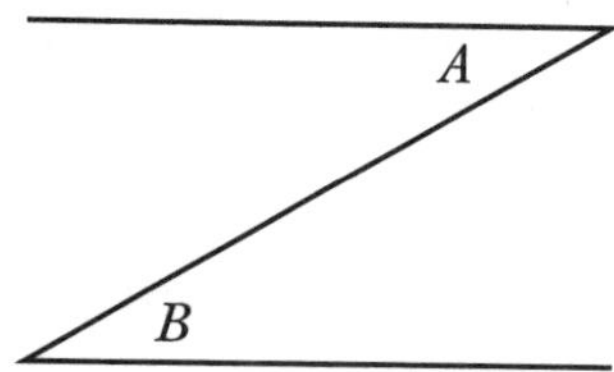

The lemma says that the two angles marked *A* and *B* are equal. While we do not intend to provide a full proof of this lemma,[20] it is worth pausing to convince ourselves that it is indeed true. Somehow, the angle *A* measures how "tilted" the top of the Z is from the diagonal. If the top of the Z were to start slanting down, the angle *A* would get smaller.

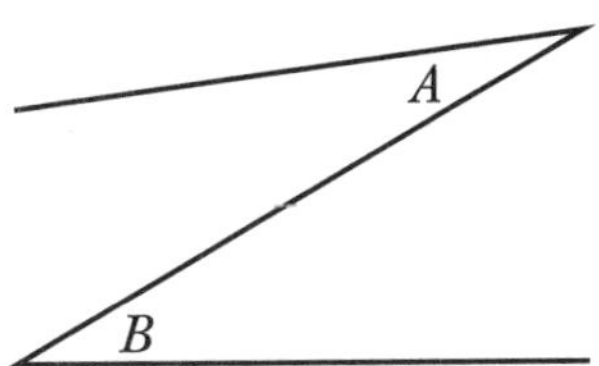

Similarly, the angle *B* measures how tilted the bottom of the Z is from the diagonal. Because the top and the bottom are parallel, their tilt from the diagonal must be the same. Again, this argument is not a formal proof, but it gives us the general idea of why the lemma is true and is enough for right now.

What, then, does the Z Lemma have to do with our Triangle Angle Sum Theorem? The task is to look for a Z in our diagram.

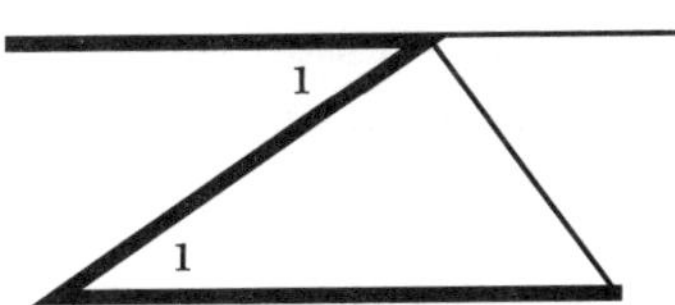

By the Z Lemma, the two angles marked "1" are identical. If we look carefully enough, we find a second Z, albeit backwards, with the angles marked "2" being identical.

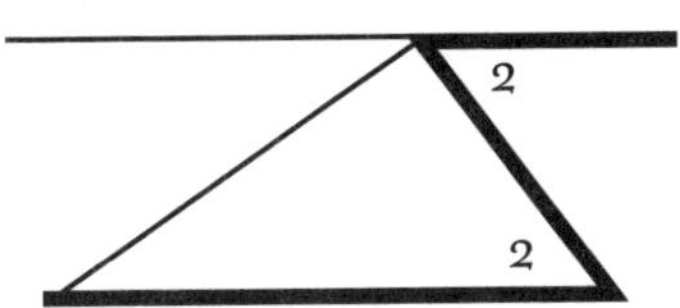

20 Euclid proved a version of the Z Lemma in which he called the angles labeled *A* and *B* alternate angles. This result is tightly linked to his famous fifth postulate, the Parallel Postulate, which we will discuss later.

We are now in a position to see that all three angles of the triangle have been rearranged to sit on a straight line.

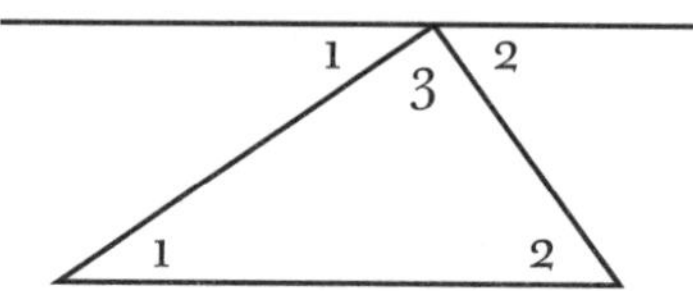

This completes our proof, but it is worth pausing to reflect and confirm that no step was tainted by the fact that our triangle was necessarily drawn imperfectly. We did not, for example, actually measure the angles with a protractor. Further, no part of the argument had anything to do with the *specific* triangle we had drawn. It is true that our triangle was acute, but if you take the time to draw a right triangle or an obtuse triangle, you will see that the argument itself doesn't change. Regardless of our starting triangle, a line can be drawn through the top vertex that is parallel to the bottom,[21] and the two Zs would be formed, allowing us to find pairs of identical angles.

Before leaving this discussion, it is helpful to introduce a similar but decidedly different proof of the same Triangle Angle Sum Theorem, if only to demonstrate that a theorem can have multiple correct and elegant proofs. In his proof, Euclid adds an auxiliary parallel line to his triangle, but in a different place.

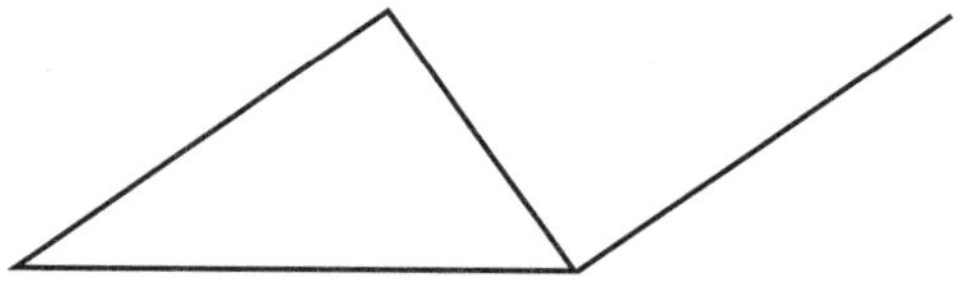

He also extends the bottom side of the triangle. The matching angles are labeled accordingly.

21 Euclid and other mathematicians go to great lengths to argue each step of a proof. The fact that a line can be drawn through a point that is parallel to another line is a proposition that Euclid deals with and successfully proves. This particular task, actually, is a source of great controversy. The discussion of how many lines can be drawn in this way given a starting point and line gives rise to the non-Euclidean geometries. When the answer fails to be 1, we also end up with non-Euclidean triangles for which the angles do not add to a straight line. While fascinating, this piece of mathematics is not necessary for understanding the basic proof presented here. We will discuss these non-Euclidean geometries later.

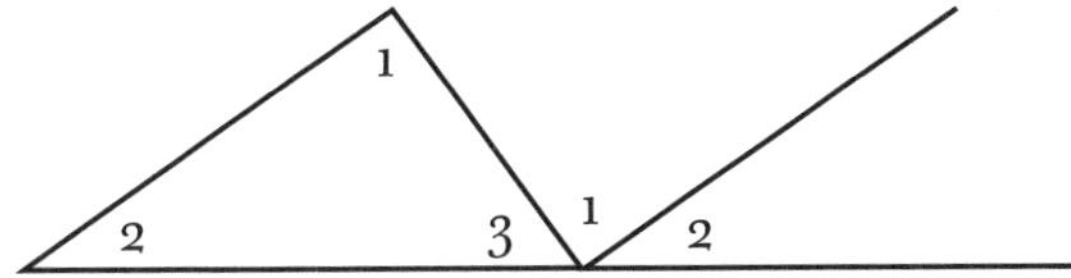

The angles marked “1” are identical because of the Z Lemma. (Can you locate the Z?) The angles marked “2” are identical, but this requires a different lemma concerning angles and parallel lines. We leave it to the reader to think through what that lemma might say.

As a final thought, we return to the theorem itself and its somewhat surprising result. No matter what triangle we use, no matter how big or how small it is, no matter how equal or unequal the sides are, no matter how small the smallest angle is, the three angles can always be rearranged to form a straight line. This is a fundamental and amazing reality of geometry and serves as a wonderful point of departure for the epic journey that lies ahead. And with that, we’re off!

PART II
TEN PROOFS EVERYONE OUGHT TO KNOW

CHAPTER 1
WHICH REGULAR POLYGONS TILE THE PLANE?

Carpet patterns, tile patterns, and even fabric patterns are often created by using the same shape repeated over the entire surface. This doesn't work with all shapes, of course, but only those that fit nicely together. It turns out that this property of "fit" can be described and explored mathematically. A shape is said to *tile the plane* if copies can be lined up in a way that leaves no gaps and creates no overlaps. Sometimes we refer to these tiling patterns as *tessellations.* As an example, consider a simple cross.

To see how it tiles the plane, we can start lining up copies of the shape. There are no gaps and no overlaps. More importantly, this pattern can continue so that it will fill a surface that is as large as we want, be it a floor, a wall, or something else.

There are numerous shapes that will tile the plane, some with lots of symmetry and some without much symmetry. Sometimes the tiling uses only one shape and sometimes more than one shape. Some of the art of M. C. Escher is based on these tiling patterns. For example, *Sky and Water* beautifully fits together the shapes of birds and fish that seem to morph into one another as our eyes move upward from the water to the sky.

Mathematicians are very interested in tiling patterns, particularly in classifying them. Some patterns look complicated but are actually based on a much simpler tiling. For example, the cross pattern is built upon a pattern of squares, which are among the simplest of shapes that will tile the plane.

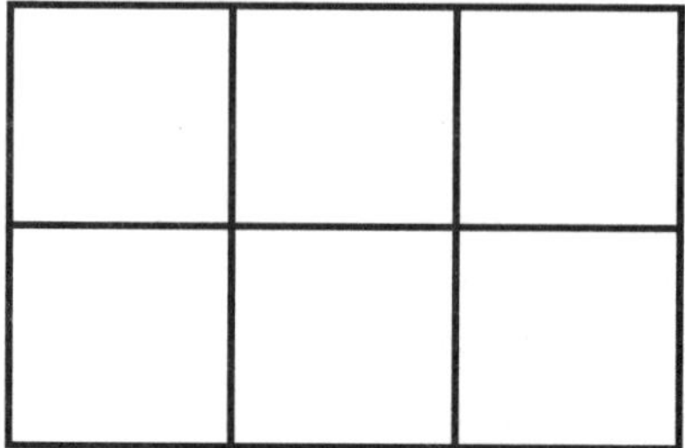

The cross pattern has a square as its basic building block.

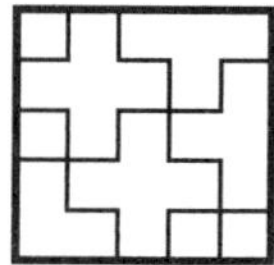

If we line these shapes up using the square pattern, we can see the full cross-based tiling.

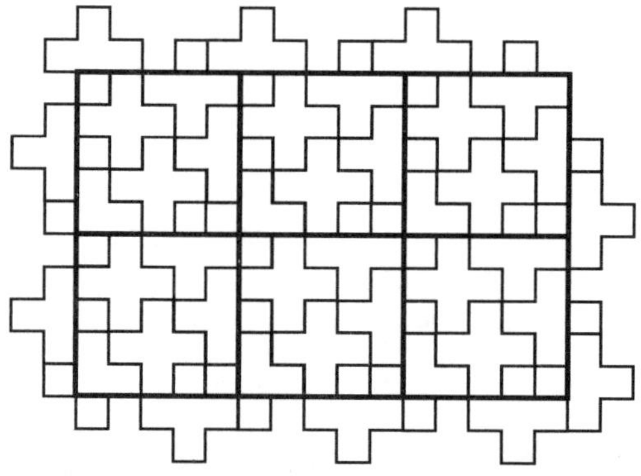

A square is an example of a *polygon*, which is a closed shape—meaning there is a clear "inside" and "outside"—formed using only straight lines that meet at the edges but do not cross otherwise. But a square is also an example of a *regular* polygon, which has the added characteristic of having all equal sides and all equal angles. The first four regular polygons are the equilateral triangle, the square, the regular pentagon, and the regular hexagon.

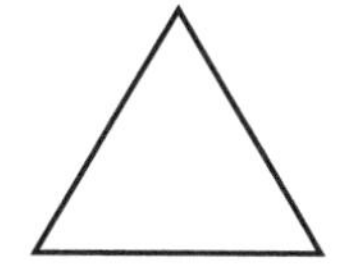

Regular Triangle (Equilateral Triangle)

Regular Quadrilateral (Square)

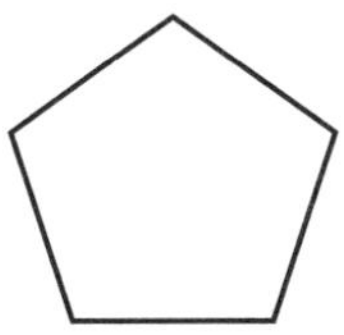
Regular Pentagon

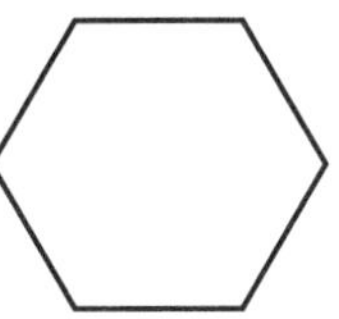
Regular Hexagon

Because of their inherent symmetry, the regular polygons are great candidates for tiling patterns. They are only candidates, however. We cannot be certain that the symmetry is enough to allow the shape to fit together nicely with copies of itself.

In the previous chapter, we started with the statement of a theorem and then looked for a proof. This was because the theorem about the angles in a triangle is a well-known result, so our task was to search for its justification. In this chapter I will start with a question rather than a theorem.

> QUESTION: Which regular polygons will tile the plane?

I start this way in part because it is possible that you, dear reader, do not yet know the answer, and to state the theorem right off the bat would give away a process of discovery. I also start this way because the process of discovery often provides us with a proof of the answer along the way. Mathematics can work in either direction. Sometimes we suspect an answer (or a theorem), and we look to prove it, whereas other times we start with a problem and do not yet know the solution.

Before we set about answering the question, we should make sure that we understand the rules of the game. We have already defined a regular polygon, and we have already introduced what it means to tile the plane. We should, however, clarify one more important condition for our quest: a regular polygon is said to tile the plane *when only exact copies of the shape are used*. In other words, it might be the case that differently sized squares can be put together in a tiling pattern, and such patterns can be beautiful and interesting. But it is not the question with which we are dealing at the moment. We are only interested in whether or not squares of exactly the same size can accomplish this feat.[22]

22 To be precise, such tilings that use exact copies of the same shape are referred to as regular tilings. There are also

Of course, in the case of squares, we have already seen that the answer is yes. But the square is not the simplest of the regular polygons. That privilege belongs to the familiar equilateral triangle. Can this honored shape align with itself in just the right way to produce a tiling pattern? The only way to know is to try. We begin by lining up one equilateral triangle with another.

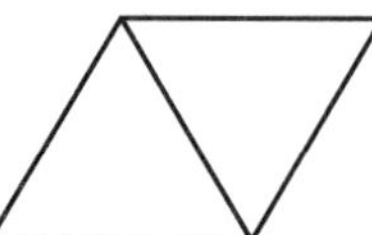

We know that the sides line up because we are dealing with two identical equilateral triangles, meaning that the sides from both triangles are all equal to one another. From here we can continue lining up triangles along these equal sides.

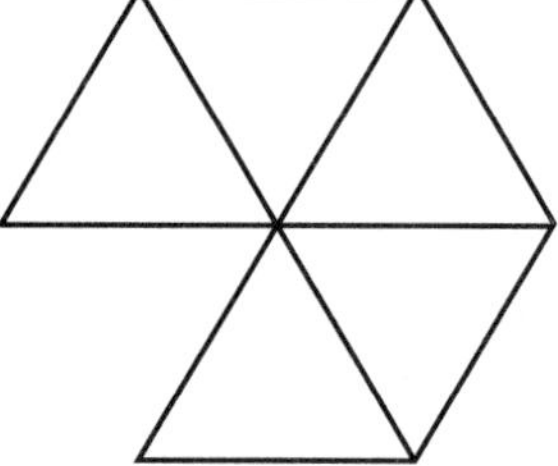

When we see these five triangles come together, we start to have hope that a tiling pattern might be possible. After all, the gap in this picture looks just big enough to fit a sixth triangle.

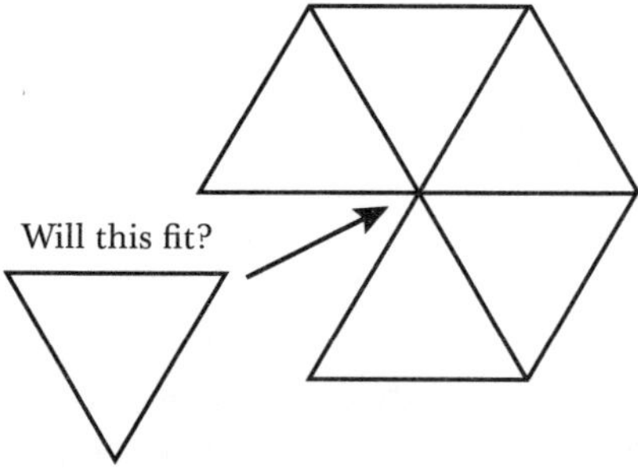

semi-regular tilings, which use the same shape but allow for different sizes, as well as irregular tilings, which allow for differently shaped tiles altogether. A particularly interesting group of irregular tilings are known as *Penrose tilings*, named for the British mathematician Roger Penrose. These irregular tilings are also aperiodic, meaning that even the way the tiles fit together isn't repeated. In other words, you can never "shift" the tiling so that it lines up with itself. These are marvelous objects.

For the mathematician, though, "looks just big enough" won't cut it. We need to be sure that when we put that sixth triangle in place we are not "fudging" the diagram. After all, to tile the plane we are looking for an *exact* fit, with no gaps and no overlaps, no matter how small. As for the sides of this sixth triangle, we know these will match up nicely. The only thing in question is how the angle fits (or doesn't).

What is the measure of one of the angles in an equilateral triangle? From our previous work on the Triangle Angle Sum Theorem, we know that the angles in any triangle can be rearranged to form a straight line, meaning that the measures of the three angles will add to 180°. The equilateral triangle, however, is no ordinary triangle. All three angles are the same size, so each of them has to be 180° ÷ 3, or 60°. Since we are dealing with identical equilateral triangles, all of the angles are 60°.

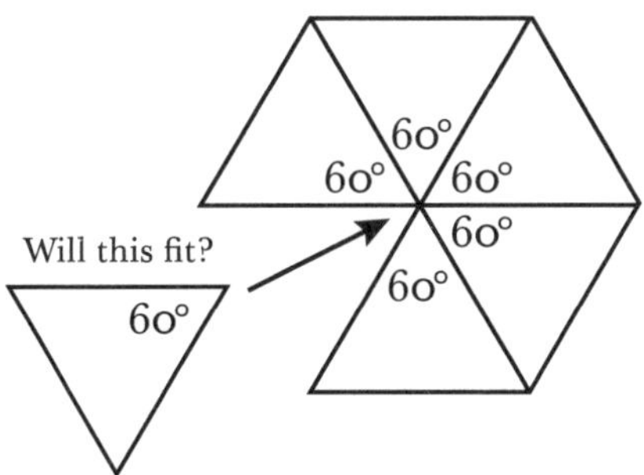

How might we figure out the size of the angle in the gap? It might help to know that "all the way around" is the same as 360°. This is because "all the way around" is the same as two straight lines.

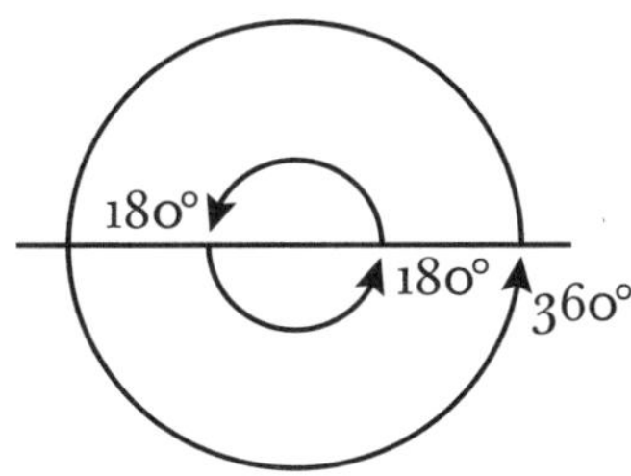

In our diagram with five triangles, we have already accounted for $5 \times 60°$, or 300°. This means there are exactly 60° left!

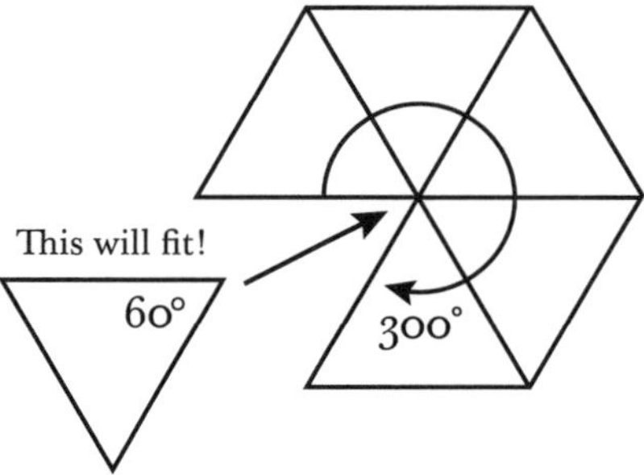

The important fact here is that when six equilateral triangles come together around a central point, the central angle that they form measures *exactly* 360°! In other words, these six triangles go all the way around the central point exactly once with no gap and no overlap.

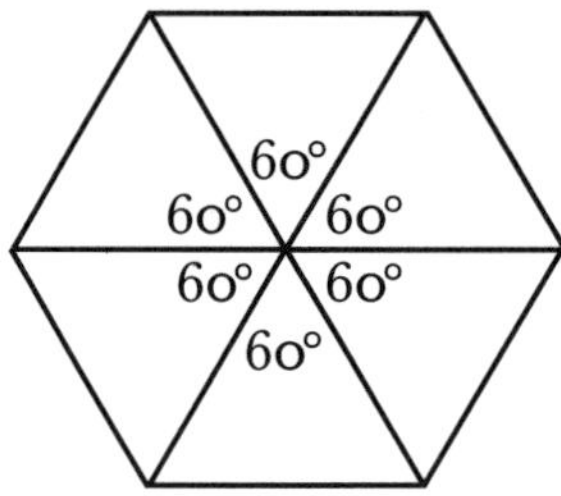

Therefore, our fit is perfect, and we can conclude that equilateral triangles will tile the plane. Once we have established this fact, we can keep lining up more and more equilateral triangles in a tiling pattern to fill as much space as we want.

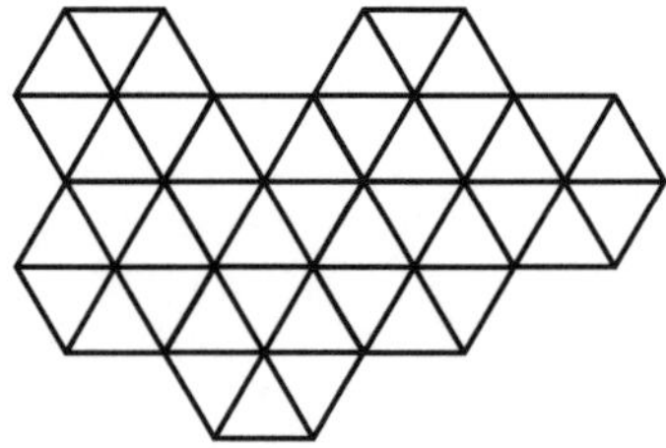

Notice that the same argument can be used to justify the fact that identical squares will tile the plane. Each angle in a square is exactly 90°, and so four squares will come together to form $4 \times 90°$, or a perfect $360°$.

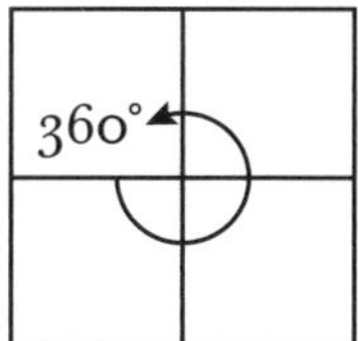

We now know that regular triangles (equilateral triangles) and regular quadrilaterals (squares) will tile the plane. At this point, we might consider our question a bit boring with two examples and two victories. After all, if *every* regular polygon tiles the plane, that might be a mildly interesting fact, but it would put our question to rest pretty quickly. Lest we be hasty, though, we should check the next case: the regular pentagon. When we try to align regular pentagons in a tiling pattern, we find that a small gap is created.

Of course, just as we were rightfully concerned about any fudging in our drawing of triangles to check that they fit, we should be concerned about the opposite here. Is there really a gap in this diagram, or were we just not very good at drawing regular pentagons? (After all, drawing regular pentagons is not an easy task. You should try it!)

In the case of equilateral triangles, we determined that each angle is 60° and reasoned from there that six of these triangles will fit nicely around a central point. In the case of squares, we knew that the measure of each angle was 90° and used this to show that four squares would accomplish the same thing. It makes sense to try the same tactic for our regular pentagons, which to begin with means finding the measure of one of its angles.

The measure of one of the angles in a regular pentagon is not well known, and cutting the angles apart and rearranging them like we did with the triangle in the previous chapter would prove to be a good bit more difficult. But when there is something that we don't know how to do, one path forward is to see whether we can relate it to something that we *do* know how to do. We spent a fair amount of time defending the fact that the angles in any triangle add to 180°. Maybe we can make our pentagon look like a collection of triangles glued together.

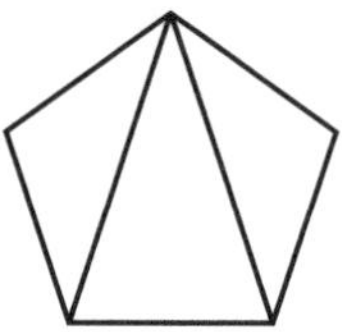

It turns out that a pentagon can be thought of as three triangles.[23] Because the angles in a triangle add to 180°, it follows that the angles in the pentagon must add to 3×180°, or 540°. Although we are dealing with a regular pentagon, this triangularization actually works for *any* pentagon. Thus, we have discovered a theorem along the way to answering our question.

> *Pentagon Sum Theorem*
> The angles in any pentagon add to 540°.

In our tiling question, however, we are dealing with a *regular* pentagon, which means that each of the five angles must be equal. Therefore, one angle in our regular pentagon must measure 540° ÷ 5, or 108°. What does this mean for a tiling pattern? Three pentagons coming together would only be 3×108°, or 324°. This is not "all the way around," which is why we see a gap. In fact, the gap is 360° − 324°, or exactly 36°, which is not nearly enough space to fit a fourth pentagon.

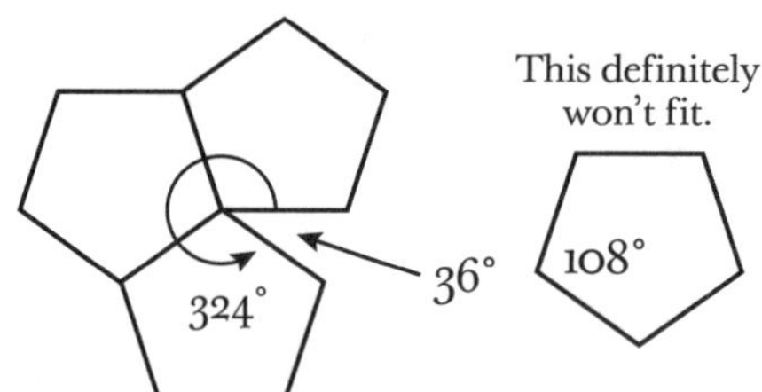

Therefore, the regular pentagon does not tile the plane.

23 I should note that the internal lines cannot cross one another in any way or create wholly new angles when doing this triangularization. It is important that all angles of all triangles can be put together to form *only* the angles of the pentagon.

Note that the same triangularization technique actually can be used to *prove* that each angle in a square is a 90° angle. I will let you draw the picture and cut the square into two triangles. The sum of the angles in a square must therefore be 2×180°, or 360°. With all four angles being equal in a square, each angle must be 360°÷4, or 90°.

This technique seems to be working, so let's move on to the regular hexagon. We start, as we did before, with finding the size of each angle. By drawing diagonals, a hexagon can be seen as four triangles.

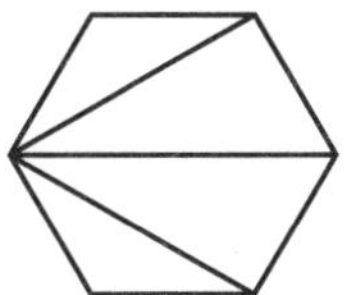

The sum of the angles in the hexagon must be equal to the sum of the angles in four triangles: 4×180°, or 720°. In a *regular* hexagon, where all six angles are equal, each angle must be 720°÷6, or 120°. Now, if each angle in the hexagon is 120°, can we arrange some regular hexagons around a central point to make it "all the way around" with no gaps and no overlaps?

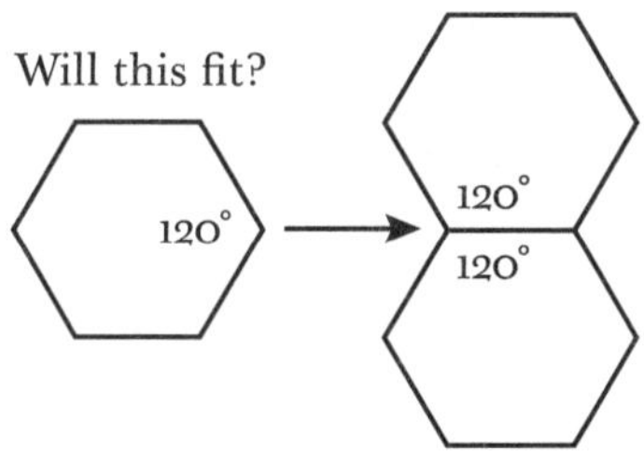

Yes! Three hexagons will do the trick because 3×120° is exactly 360°. Therefore, three hexagons will come together to tile perfectly. This should not surprise us, actually. If we continue this tiling, we find the familiar "honeycomb pattern" that appears frequently in nature.

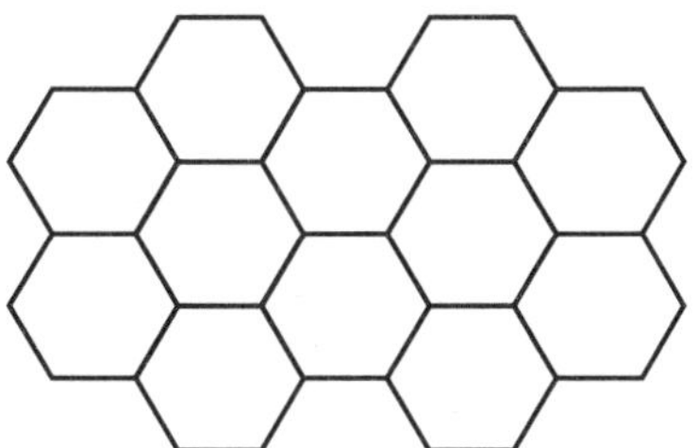

In fact, if we were perceptive, we should have seen that this hexagonal tiling is related to the tiling pattern with equilateral triangles.

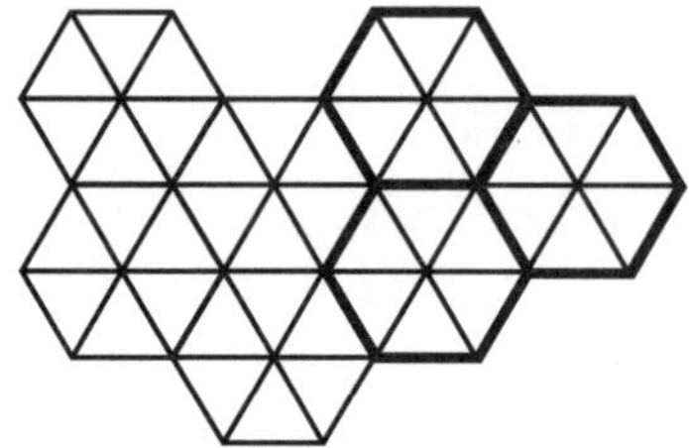

At this point, it is worth pausing to review what we know.

Name	Number of Sides	Sum of the Angles	Measure of One Angle	How Many Shapes Come Together	Does It Tile?
Equilateral Triangle	3	180°	60°	6	Yes
Square	4	360°	90°	4	Yes
Regular Pentagon	5	540°	108°	3 (but there is a gap)	No
Regular Hexagon	6	720°	120°	3	Yes

We have done a lot of work, but we have only dealt with four shapes. (Challenge yourself to fill out the six columns for a regular 100-gon, i.e., a regular polygon with 100 sides, to see whether it will tile the plane.) Have no fear: we only have an infinite number left to check! In all seriousness, we certainly do not have the time to check an infinite number of cases. How, then, ought we to proceed? Mathematics again comes to the rescue with its ability to reason universally about a property of an infinite number of things, in this case regular polygons.

The chart we have created paves a path forward. What patterns do you see in the chart? Specifically, as we move down the chart and the number of sides increases, what happens to each of the other columns? The numbers in the "Sum of the Angles" column are increasing. This makes sense, right? The sum of the angles came from counting the number of triangles in the triangularization of the regular polygon. As a side is added—for example, moving from a regular pentagon to a regular hexagon—exactly one more triangle is needed. Therefore, the sum of the angles goes up by exactly one triangle, or 180°. The numbers in the "Measure of One Angle" column are also increasing, which should also come as no surprise. As more sides are added to the regular polygon, the angles have to be "wider" to fit all these sides. On the other hand, the numbers in the "How Many Shapes Come Together" column are *decreasing*. Why is this? As the angles in the shape get wider, we cannot fit as many shapes "all the way around" in the quest to find a perfect 360°.

It is this last column that contains the clearest path forward for answering our question. If these values are decreasing, and if the number of shapes for the regular hexagon is already 3, where can they go from here? These numbers must be whole numbers because they represent a number of shapes. In fact, in the case of the pentagon the number was *not* a whole number—it was between 3 and 4—which is precisely why the pentagon does not form a tiling pattern. Using only whole numbers and counting *down* from 3, we can only go to 2 or 1. Therefore, if a regular heptagon, octagon, nonagon, or any other shape after the regular hexagon were to tile the plane, the number of shapes coming together around a central point would have to be either 2 or 1. The case of 1 makes no sense at all. The whole point of tiling is that *more than one* copy of the shape are coming together to form a perfect fit.

What about the case of 2? What would it mean for exactly two copies of a shape to come together at a central point? If "all the way around" is 360°, then each of the two angles would have to be 360° ÷ 2, or 180°. The angle in the regular polygon would have to be 180°, but 180° is a straight line. This is impossible because a single angle in a polygon, regular or otherwise, cannot be a straight line.

Therefore, we have answered our question in perhaps a surprising way. When we first set out to discover which of the infinite number of regular polygons can be used to tile the plane, we might have thought that an infinite number of them would rise to the occasion. In fact, after confirming that the equilateral triangle and the square work, we briefly paused to consider that maybe *all* of them might work. Even after discovering one that didn't work (the regular pentagon), we might have still thought that an infinite number of cases could work. Maybe only multiples of 5 cause a problem. Maybe *only* the case of 5 causes a problem. Maybe only multiples of 2 and 3 work and everything else is a problem. In any of these situations, we would still have an infinite number of regular polygons that tile the plane.

This is not the case, however. What we have discovered, perhaps surprisingly, is that only three cases work: the equilateral triangle, the square, and the regular hexagon. This is our theorem, and our process of exploration has provided us with a proof along the way.

> *Regular Polygon Tiling Theorem*
> There are only three regular polygons that tile the plane: the equilateral triangle, the square, and the regular hexagon.

And once again, mathematics has been able to deal with an infinite number of cases without actually checking each of the cases individually.

CHAPTER 2
THE PLATONIC SOLIDS

THE REGULAR polygonal tiling patterns provided us with a surprising and beautiful result: only three of the infinite number of regular polygons will tile the plane: the equilateral triangle, the square, and the regular hexagon. The analogous question in three dimensions is "Which regular three-dimensional polygon-type objects will tile space?" As you might imagine, "regular three-dimensional polygon-type objects" is not really a mathematical term, so we have some work to do to figure out what a three-dimensional version of a regular polygon might be.

To begin with, we insist that all of the faces of our shape be polygons. These types of solids are known as *polyhedra*, which are the three-dimensional analog to polygons. Next, we define a *regular polyhedron* as one with faces that are identical regular polygons *and* has the same number of polygons coming together at the points, or *vertices*. This second condition of regularity is important. Consider the following shape, which is made with ten congruent equilateral triangles.

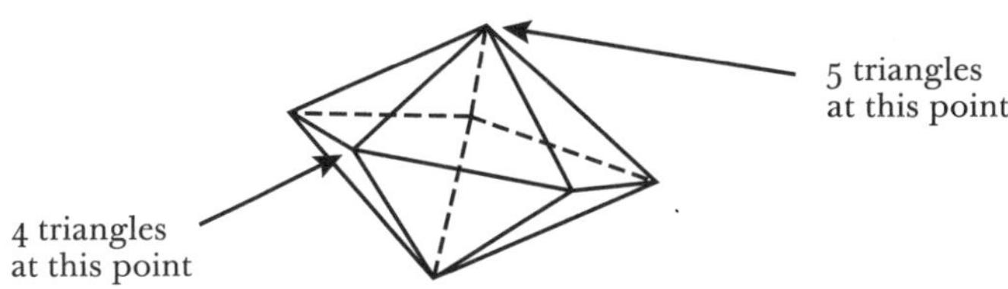

All the faces might be identical and regular, but at some points (the top and the bottom) there are five triangles coming together to form the vertex, whereas at other points (along the sides) there are only four triangles coming together. Therefore, while certainly a polyhedron, and one with a lot of symmetry because of the equilateral triangles, this shape is not a *regular* polyhedron.

But even with the condition of having the same number of polygons coming together at each vertex, we do not yet have a full definition of the sort of shape in which we are interested. We really need to keep our shape from caving in. To see why, let's back up again to two dimensions and consider our old tiling friend, the cross.

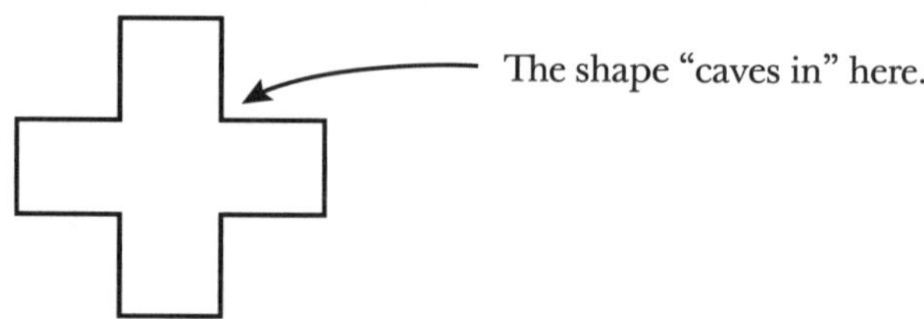

In mathematical parlance, shapes that cave in are aptly called *concave.* What does this have to do with being regular or irregular? Notice that all of the sides of the cross are equal, but it is certainly not a regular polygon. In two dimensions, we know this because the angles themselves are not equal. Some of them are 90°, and others are 270°.

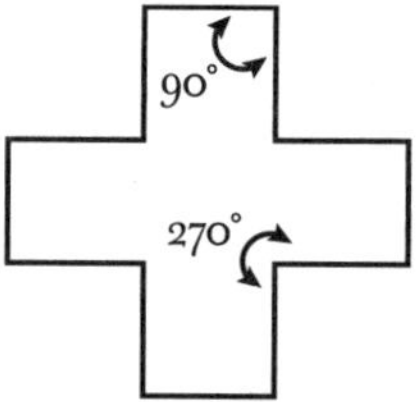

Angles in three dimensions are going to be quite a bit trickier to measure, and we would rather avoid that complexity. Instead, we will insist that our polyhedra are *convex*. Convex is the opposite of concave, and so a convex shape does not cave in. More precisely, being convex means that any two points in the shape can be joined with a straight line that does not leave the shape. The cross is not convex because there are pairs of points for which the straight line joining them temporarily leaves the shape.

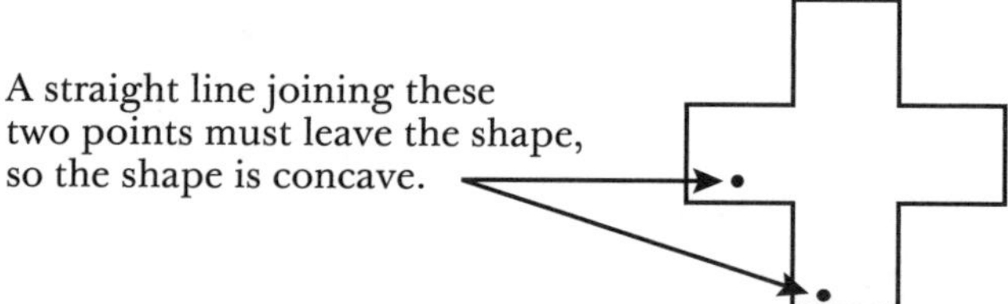

The same definition can be applied to polyhedra, and just like in two dimensions, three-dimensional shapes can be either convex or concave. The difference is that regular polygons are all convex, but regular polyhedra can be either convex or concave. As an example, the *stellated octahedron* is formed with thirty-two identical equilateral triangles and three triangles joining together at every vertex.[24] The faces, however, cave in along certain edges.

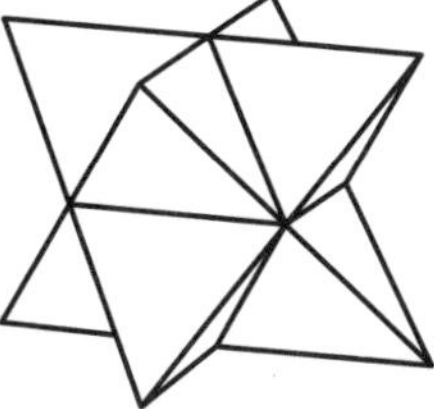

As a concave shape, the stellated octahedron is not quite symmetric enough for our purposes. The shapes in which we are *really* interested, those that are three-dimensional analogs of the regular polygons, are *convex regular polyhedra*. These "super regular" shapes are known as *Platonic solids*. In the *Timaeus*, Plato discusses the nature of the physical world and along the way mentions the existence of these regular three-dimensional shapes, even going so far as to associate them with what he thought were the basic elements that make up the universe. I beg your patience for a moment for a short digression into a bit of art history and philosophy. In Rafael's famous painting *The School of Athens*, Plato and Aristotle are featured prominently in the center. There are two notable differences between these philosophers in the painting. First, Plato points upward, likely indicating his position that we come to know reality through ideals, an idea that emphasizes our earlier observation that the perfect triangle is not something we have ever seen but rather something that only exists in the heavens. Aristotle points downward, likely indicating his support for the idea that we know reality primarily in the things of the earth, which emphasizes our other observation that the material universe speaks of mathematical realities.[25] The second difference is the book held by each philosopher. Plato is holding none other than the *Timaeus*. A work that describes time, space, and change, it was used for hundreds of years as something of a manual for mathematics and physics. Aristotle is holding a copy of the *Nicomachean Ethics*. This might be significant because at the start of his text Aristotle strongly holds that the study of ethics cannot be reduced to or known with the same certainty as mathematics.

24 The definitions of regularity can vary from source to source, and the best sources will make distinctions based on symmetry that we are not going into here. Our only interest is in avoiding complicated ways in which angles are measured in three-dimensional shapes. Officially, the stellated octahedron is a "regular polyhedral compound."

25 Yes, this is an oversimplification. Tomes have been written on how these two philosophers understand ultimate reality, how we experience it, and how we come to know it.

At any rate, because Plato discusses these solids in the *Timaeus*, they have become known as the Platonic solids. We will have more to say about this later, but returning from our sidebar, we recap with a definition.

> DEFINITION: A *Platonic solid* is a three-dimensional shape with the following properties:
> 1. The faces are identical regular polygons.
> 2. There are the same number of regular polygons coming together at each vertex.
> 3. The shape is convex.

The most well-known Platonic solid is the cube, formed with six identical squares.

It also turns out that this Platonic solid will tile space.

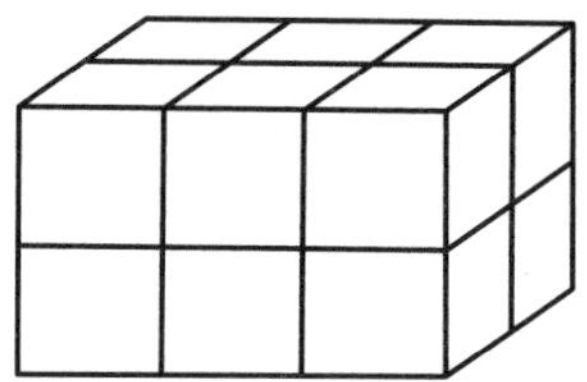

When we explored the question of tiling the plane, we began by noting that there are an infinite number of regular polygons that serve as candidates. The problem with the Platonic solids, the basic building blocks of our three-dimensional tiling patterns, is that it is not clear what they are, what they look like, or even how many there are. We only have one example thus far: the cube.

To set out on a quest to discover more of these objects, we start with the second of the three properties in the definition: having the same number of regular polygons joining together at a vertex. Let's look carefully at the vertex of the cube, but let's do so by unfolding it. We are interested in the center point at which *three* squares join together. As the shape is refolded, this center point becomes a vertex of the cube.

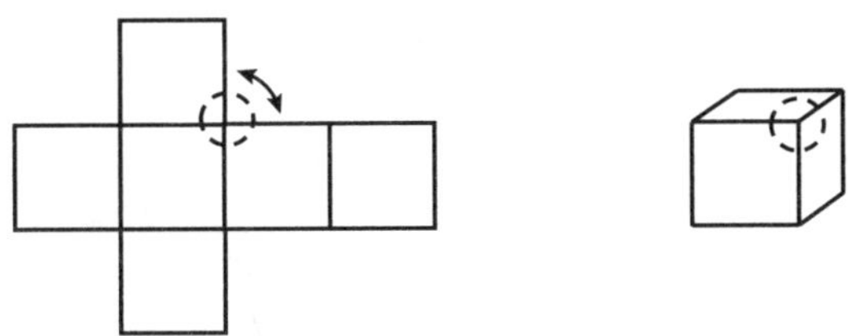

Notice the similarity and difference between this arrangement of the squares around this point and the arrangement that we used to tile the plane with squares.

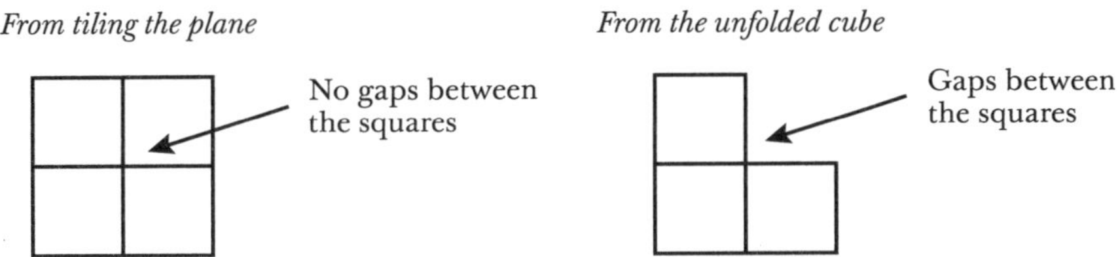

In tiling the plane, we wanted no gaps and no overlaps between the squares. In this case, however, because the squares are going to be folded up into a three-dimensional shape to form the cube, we *need* a gap. We can think about forming this portion of the cube either by folding up the two squares so that the edges come together or by pulling up on the center point. Either way, the two edges will then align, forming the vertex of the cube.

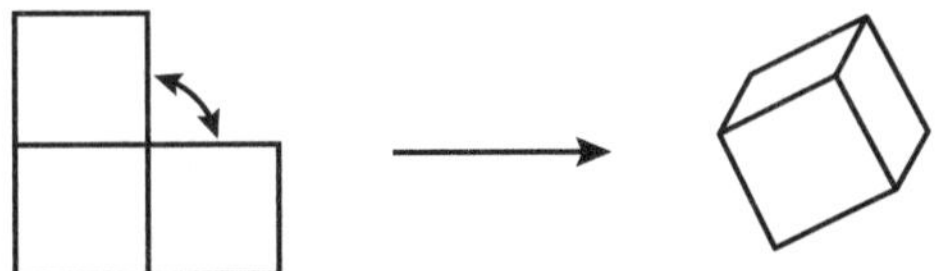

We need the gap between squares, otherwise we cannot bring the unfolded shape into three dimensions. In other words, unlike in tiling the plane, when we wanted the angles to come together to make *exactly* 360°, here we need the squares to come together to be *less than* 360°. The observation of a necessary gap is significant, and it puts us in a position to start thinking about how to construct other Platonic solids.

Like we did with regular polygons, let's move backwards from squares to equilateral triangles. What sorts of three-dimensional shapes can be made using these? You will remember from the previous chapter that one angle in an equilateral triangle must measure 60°. In thinking about how the combination of these triangles could be folded or lifted to form the vertex of a solid, it's worth noting that we need *at least three* of them around a central point. Why? It seems clear that *one* triangle cannot be used—after all, there would be nothing to fold or lift—but what about *two*? The two triangles could of course be folded up, but they would not come together to form a solid. Rather, they would just lie on top of one another to form a "doubly thick" triangle, though "doubly thick" does not really have meaning given that the triangles have no thickness to begin with.

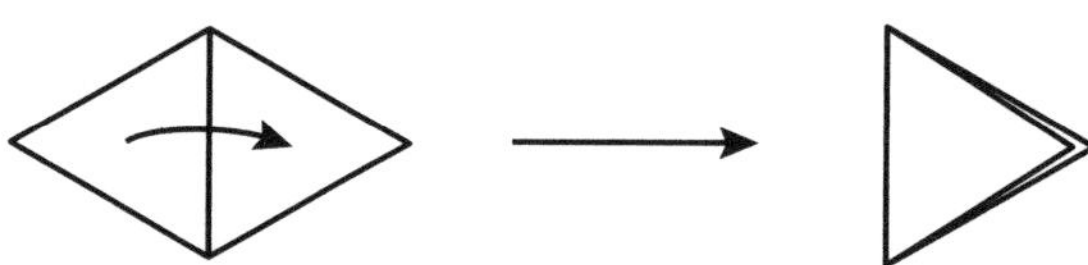

Therefore, the minimum number of triangles we need is *three*. The following diagrams represent the bringing together of three, four, five, and six equilateral triangles.

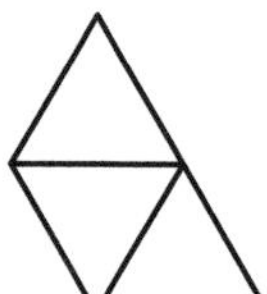

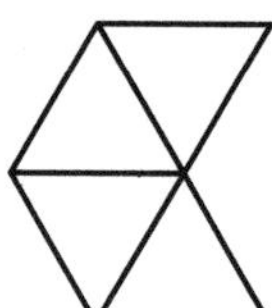

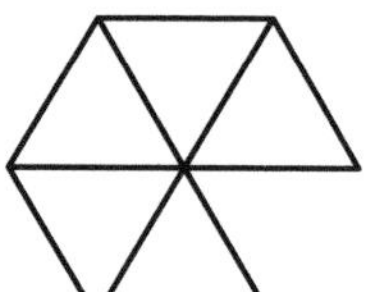

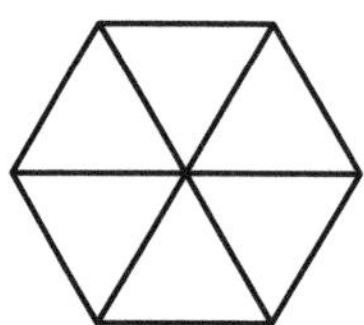

Remember, we need a gap in order to account for the folding or lifting of the vertex to bring the shape into three dimensions. The first three have a gap that can serve this purpose, but the last does not. In fact, the last is the tiling pattern from our previous work in which six triangles fit together perfectly without gap or overlap. Therefore, only the first three have the potential of forming a Platonic solid. Moreover, we need not consider using any more triangles to form a vertex. Any more triangles will overlap rather than leave the needed gap. This means that there are at most three different Platonic solids using equilateral triangles. We will consider each of them in turn.

The case of three triangles coming together at a vertex would fold up nicely.

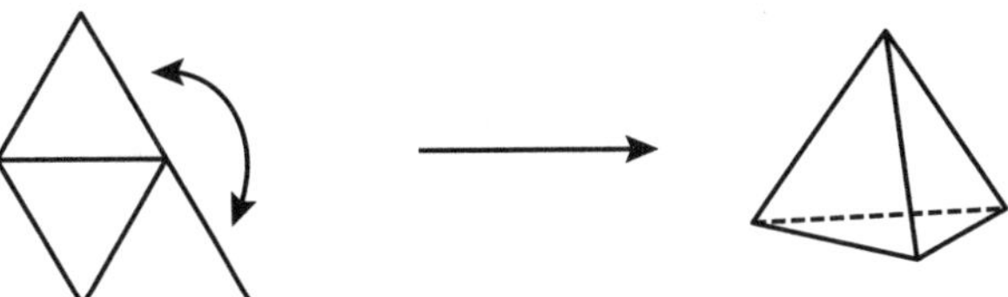

The result is a pyramid that has an equilateral triangle as a base. In fact, we could have seen that the base would end up as a triangle by counting the sides that are *not* coming together.

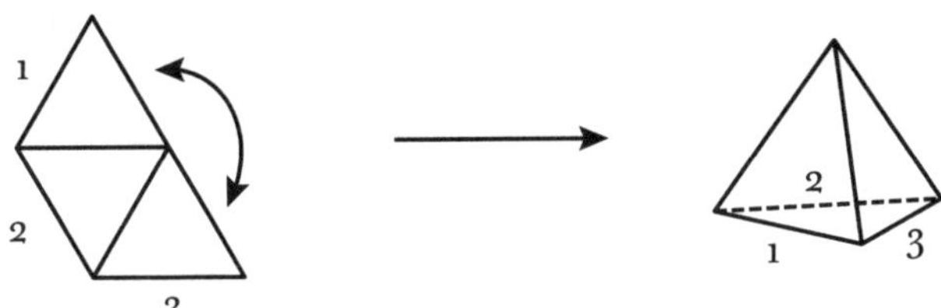

Because the edges of the pyramid's base (the sides marked 1, 2, and 3) come from the original three triangles, themselves equilateral, the base *must* be an identical equilateral triangle. Therefore, because all four faces are the same, we have found our second Platonic solid. The name of this solid is a *tetrahedron*.

We move on to the case of folding up four triangles. The result will be a pyramid with a square base. Again, we know the base will have four equal sides because there are four equilateral triangles coming together.

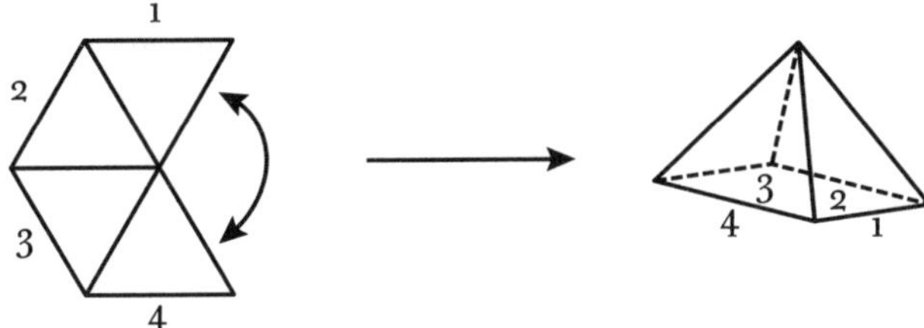

This pyramid is *not* a Platonic solid because the base, which is a square, is not the same shape as the other faces. But we can double the shape and glue it to the bottom, and the result is the Platonic solid known as the *octahedron*.

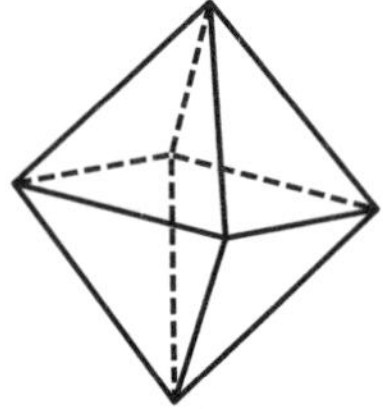

We should check carefully that this shape satisfies the three properties of a Platonic solid. First, all sides are identical equilateral triangles. Additionally, it is convex. But do we have the same number of shapes joining together at each vertex? The answer is yes. We already know that the top and bottom vertices are formed by four triangles joining together. Around the sides, you can count the four triangles at each vertex. Therefore, the octahedron is the third Platonic solid.

Finally, for triangles at least, we consider five triangles being pulled up into a vertex, which should leave a base of a regular pentagon.

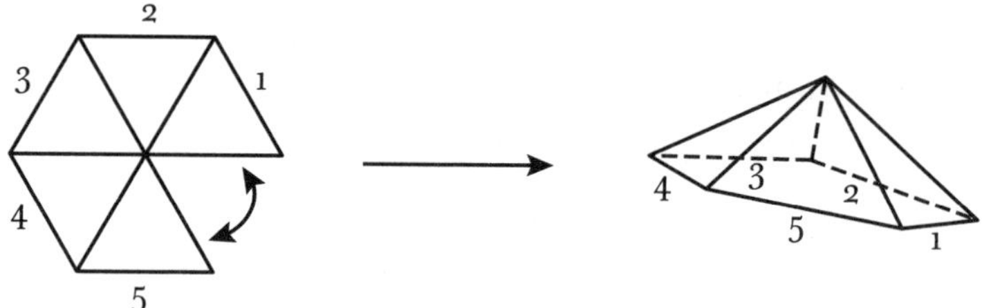

As before, this shape is not a Platonic solid because the base is a pentagon, and the other faces are triangles. We might, however, try to combine two copies of this shape in the same way we did to form the tetrahedron.

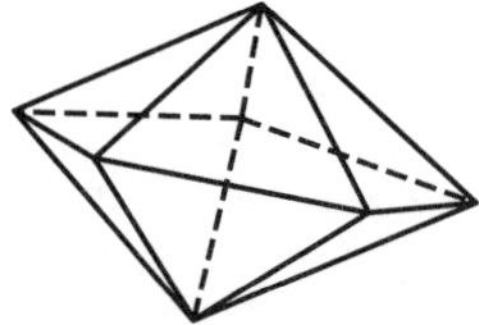

We saw this shape earlier in the chapter. While highly symmetric, it is not a Platonic solid because there are not the same number of triangles joining together at each vertex. Nevertheless, all is not lost. It turns out we *can* replicate copies of our "pentagonal pyramid" using a more ingenious construction. It takes twenty triangles, so four copies of the pyramid. The result is a Platonic solid known as the *icosahedron*. It is worth verifying that there are five triangles at each vertex and worth challenging yourself to see whether you can identify the four pentagonal pyramids that make up this shape.

This is it, then, for triangles. Using equilateral triangles, there are only three Platonic solids: the tetrahedron, the octahedron, and the icosahedron.

What about using squares? As with any starting shape, we have to have at least three squares joining together to form a vertex, and we have already seen that this combination makes a cube. If we try to put four squares together, we get the tiling pattern, hence no gap that leaves room for pulling up the vertex to form a solid. Having more than four squares only exacerbates that problem. Therefore, the only Platonic solid that can be formed with squares is the cube.

Next, we consider using regular pentagons as faces for a polyhedron. We know that we need at least three starting shapes, and we have already seen what happens when you align three regular pentagons around a central point. A pesky gap was created that caused an issue with tiling. This time, however, it is just what the doctor ordered. It is the gap that allows us to pull up on the central point to get the three pentagons to form a vertex.

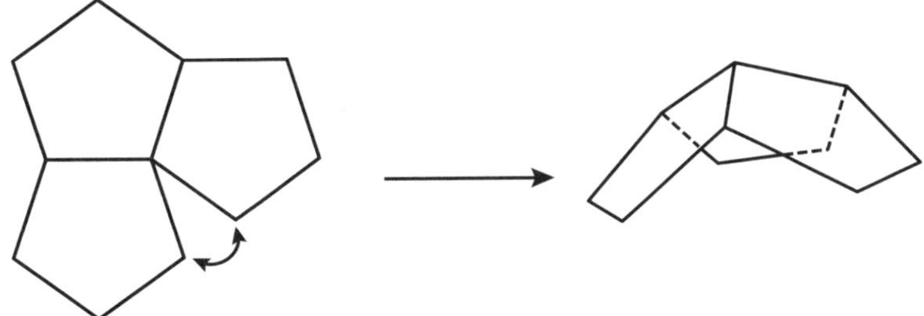

This shape is not quite a pyramid because pentagons do not act like triangles, which always "pull up" to be pyramids. As with the icosahedron, however, through an ingenious arrangement, we can combine multiple copies of this shape to make a solid. In this case, four copies create twelve identical regular pentagons that form a Platonic solid called the *dodecahedron*. At each vertex there are three regular pentagons, making twelve faces altogether.

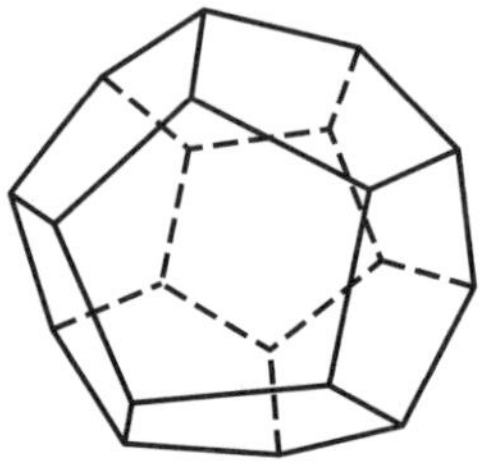

Starting with three pentagons makes this beautiful shape, but what about starting with four pentagons around a central point? The problem here is that there is no gap when it is drawn in two dimensions. In fact, we know from our previous work that the gap formed with three regular pentagons measures 36°, not nearly enough to fit a fourth pentagon with an angle of 108°. A fourth pentagon will create an overlap, so they cannot be pulled up into a three-dimensional shape. After four, the problem only gets worse. Therefore, there is only one Platonic solid that can be formed with pentagons: the dodecahedron.

Finally, we consider hexagons. Knowing that we need at least three coming together around a point, we start there.

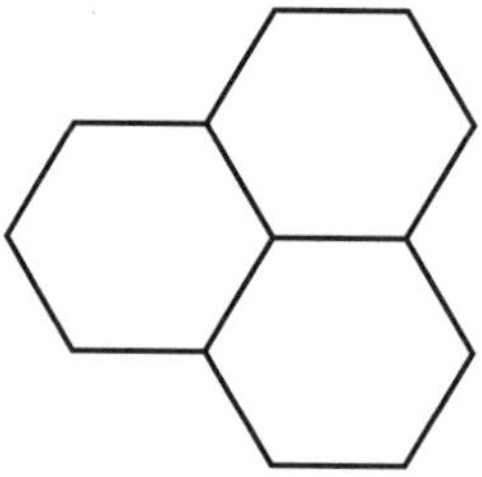

Recall, though, that these three hexagons fit together perfectly, with each central angle being 120°. There is no gap, so we cannot lift up on the central point to form a solid. More than three hexagons will cause an overlap, which is a worse problem. Therefore, we have the astonishing fact that there are *no* Platonic solids formed using regular hexagons, and yet *this was a shape that tiled the plane!* In fact, our conclusion is more astonishing than that. From this point on, the angles of the regular polygons will get larger and larger, so the minimum of three copies will always produce an overlap rather than leaving a useful gap. Therefore, there are *no more examples*. We have come to the end of our quest for these perfect shapes: there are only five Platonic solids.

Name	Face	Number of Faces	Number of Faces at a Vertex	Diagram
Tetrahedron	Equilateral Triangle	4	3	
Octahedron	Equilateral Triangle	8	4	
Icosahedron	Equilateral Triangle	20	5	
Cube	Square	6	3	
Dodecahedron	Regular Pentagon	12	3	

This result is truly astounding. Whereas there are only a finite number of regular polygons that tile the plane, there were an infinite number of these regular polygons to serve as candidates. In three dimensions, before we even consider the question of tiling, *there are only five of these perfect shapes to begin with!* In fact, it is so incredible that we consider it to be *the* theorem of this chapter, even though we first approached it in the context of the broader question of three-dimensional tiling.

Platonic Solids Theorem
There are only five Platonic solids: the tetrahedron, the octahedron, the icosahedron, the cube, and the dodecahedron.

To be precise, we actually only proved that there are *at most* these five.

Even though we explained how they came about and sketched the diagram for each, we did not actually demonstrate that the diagrams are not fudged. In order to complete this proof, we would have to actually construct each of these and show that the shapes line up the way we claim. This might be convincing enough in the case of the cube and the tetrahedron, and perhaps even in the octahedron. But in piecing together four copies of the basic pulled-up shape for each of the icosahedron and dodecahedron, I even referred to the way in which the pieces fit together as "ingenious" without demonstrating that it actually works. Euclid's *Elements* culminates in the marvelous construction of all five of these solids in Book XIII. The constructions are followed by the above proof that there are no more such solids beyond these five, and together this set of propositions forms the end of the *Elements*.[26] This proof is even more compelling when we consider that the very first proposition of Book I in the *Elements* is the construction of the equilateral triangle. It is not a stretch to say that the *Elements* starts with the construction of regular polygons and has the construction of the Platonic solids as its ultimate goal. Shapes with this amount of symmetry were that important to the ancient Greeks. As I mentioned, in the *Timaeus* Plato associates each solid with a basic element: the tetrahedron with fire, the octahedron with air, the icosahedron with water, and the cube with the earth. He then claims that the dodecahedron was used by the gods to arrange the constellations.

I have not forgotten the original question about using these shapes to tile space. At the start, we might have guessed there were only a finite number that might work. After all, that was the case in two-dimensional tilings of the plane. What we did *not* expect is that there are only a finite number of candidates to consider. We have already seen that the cube tiles space.[27] While we will not prove it here, looking at the basic shape of the other four Platonic solids will informally confirm that none of them can be lined up to tile space. Therefore, the three-dimensional version of the tiling theorem is considerably simpler.

26 Propositions 13–17 in Book XIII provide, in order, the construction of the tetrahedron, octahedron, cube, icosahedron, and dodecahedron. The proof that the five constructed Platonic solids are the only such solids comes as a remark after the final proposition in Book XIII, which is about the ratios of the sides in each solid to the diameter of the sphere in which they are inscribed.

27 There is something elegant about the fact that the cube represents the earth and that it is the only shape that tiles. It is almost like it really is a "block" from which solid matter is made. This might be a stretch, but it is an elegant one.

> *Platonic Solids Tiling Theorem*
> The cube is the only Platonic solid that will tile space.

I would be remiss if I did not end this chapter by mentioning the way in which the Platonic solids are related to one another beyond their shared regularity. If we take the very middle point on each face of the cube and use these points as vertices to form another solid, we get the octahedron. Conversely, if we take the very middle point of each face of the octahedron and use them as vertices for a new solid, we get a cube. The octahedron and the cube are known as *duals* of one another. (Whether it makes sense that air and earth go together, I am not sure.) The icosahedron and the dodecahedron are also duals of one another. (Again, whether water and the constellations pair well is not for me to say.) The tetrahedron is a dual with itself, meaning that if you take the middle point of all the faces on a tetrahedron and use them as vertices for another solid, that solid is another tetrahedron.

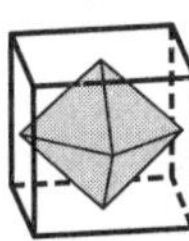
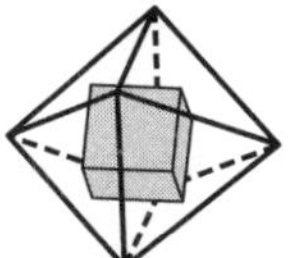
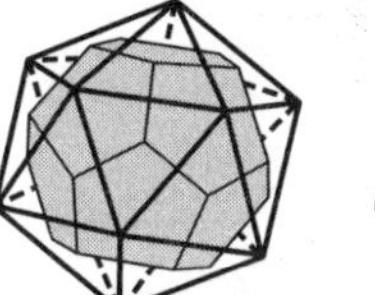
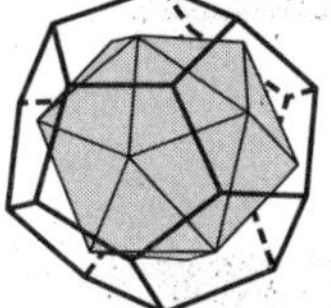

Another way to describe this duality is in terms of the number of faces, edges, and vertices in these solids.[28]

Name	Number of Faces	Number of Vertices	Number of Edges
Tetrahedron	4	4	6
Cube	6	8	12
Octahedron	8	6	12
Icosahedron	20	12	30
Dodecahedron	12	20	30

28 There is a very famous and elegant theorem by the mathematician Leonard Euler that says that in *any convex polyhedron* (not just Platonic solids), it is always the case that $V-E+F=2$, where V, E, and F are the number of vertices, edges, and faces. You can confirm this equation by using the values in the table, e.g., for the icosahedron, $V-E+F=12-30+20=2$. Perhaps Euler's formula is a good candidate for a second book.

Notice that a shape and its dual have the same number of edges. Further, notice that the number of faces is swapped with the number of vertices between duals, e.g., the cube has six faces and eight vertices, while the octahedron has eight faces and six vertices. While the basic elements of the universe and their relationship to these solids may not be quite as Plato described, it is certainly the case that they are fascinating objects worthy of an encounter.

CHAPTER 3
THE PYTHAGOREAN THEOREM

THE PYTHAGOREAN THEOREM is another theorem so famous that it is worth stating up front before setting out on a journey in search of a proof. Note, though, that its familiarity lies more in its algebraic conclusion than anything. Indeed, the equation $a^2+b^2=c^2$ is so well known that it is often mistaken for the theorem itself by those who forget that these variables don't represent just any numbers but parts of a right triangle.

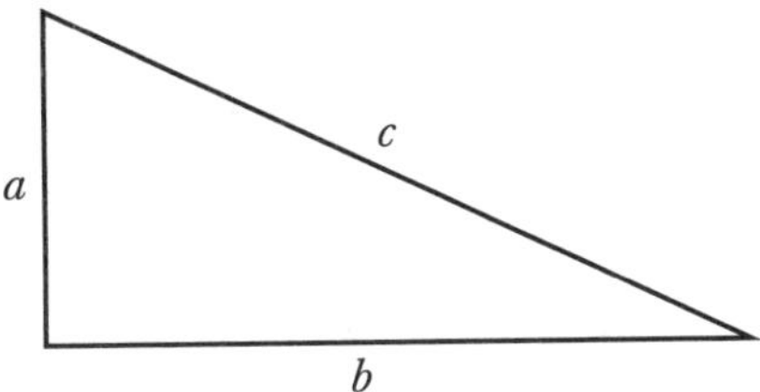

> *The Pythagorean Theorem*
> In any right triangle, the square of the hypotenuse is equal to the sum of the squares of the two legs.

Even though the modern formulation involves the famous equation, the Pythagorean Theorem is actually a geometric theorem about areas. This is how it is described at the end of Book I in Euclid's *Elements*. In fact, it is clearly the case that Book I is leading towards this result; it is the culmination of all the work that precedes it.[29] In Proposition 47, Euclid states, "In right-angled triangles the square on the side subtending the right angle is equal to the squares on the sides containing the right angle." (He leaves out the word "area," but this is what he means, as is evident from the proof itself.) The real picture of the theorem shows the squares built upon the three sides of a right triangle.

---

29 Book I actually concludes with Proposition 48, which is the *converse* of the Pythagorean Theorem. It states that if the sides of a triangle satisfy $a^2+b^2=c^2$, then the triangle must be a right triangle. Together, the two propositions show that the Pythagorean relationship holds for all right triangles and *only* for right triangles. We will not present the proof of the converse in this chapter.

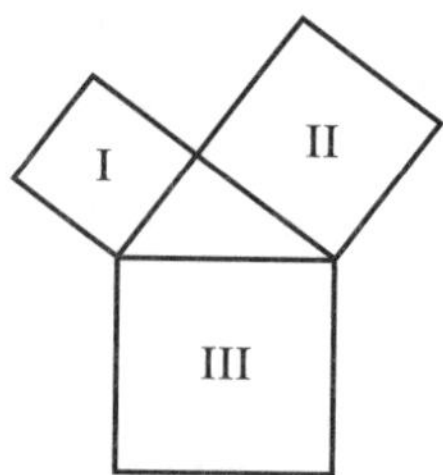

The right triangle is in the middle, and the theorem assures us that area III is equal to the sum of areas I and II. What is it about right triangles that makes this true? The theorem is not merely surprising; it is also an extremely powerful result within mathematics. (Surprising theorems are often powerful results.) While the theorem is about areas, its practical power is in what it says about lengths. When the equation is written by solving for c, we get $c=\sqrt{a^2+b^2}$, which effectively allows us to calculate the length of the hypotenuse of any right triangle so long as we know the lengths of the legs. The importance of this for finding distances cannot be overstated.

While the theorem belongs to Pythagoras, the proof we offer is Euclid's. Regarding both of these mathematicians, the fifth-century Greek philosopher Proclus wrote:

> *If we listen to those who like to record antiquities, we shall find them attributing this theorem to Pythagoras and saying that he sacrificed an ox on its discovery. For my part, though I marvel at those who first noted the truth of this theorem, I admire more the author of the Elements, not only for the very lucid proof by which he made it fast, but also because in the sixth book he laid hold of a theorem even more general than this and secured it by irrefutable scientific arguments.*

We will return to Euclid's "even more general" theorem at the end of this chapter. For now, we proceed to Euclid's "very lucid proof" of the original theorem. The reader will note that this proof takes quite a bit more work than the previous three proofs, but it is worth the effort.[30]

30 As a humorous aside, there is a proposition near the beginning of the *Elements* that is notoriously the first truly difficult one for the reader. Proposition 5 in Book I says that the base angles in an isosceles triangle are congruent. This particular proposition is referred to as the *pons asinorum*, Latin for "bridge of asses." The bridge separates the simple beast from rational man. If a reader has struggled through the proof and come out on the other side with understanding, then that person has crossed the bridge and joined the ranks of reason. The diagram for the proposition looks something like a bridge, so this provides another, perhaps less entertaining, source for its name. The Pythagorean Theorem serves a similar purpose for us. The first proof was a warm-up, and the next two gave us something of a leisurely start on our journey. It is now time to cross our own *pons asinorum*.

Most proofs have a "crux" that provides a central narrative. For the Triangle Angle Sum Theorem, the crux was to rearrange the angles onto a straight line, and then to set about showing the equality between these angles and their original counterparts. In our work on tiling the plane, the crux was to examine the angles of the polygons and how they fit or don't fit with one another when arranged around a central point. For Euclid's proof of the Pythagorean Theorem, the crux is to divide the largest square in such a way that the two pieces have areas that are equal to the areas of the two smaller squares. He does this by dropping a vertical line from the right angle to the bottom of the largest square.

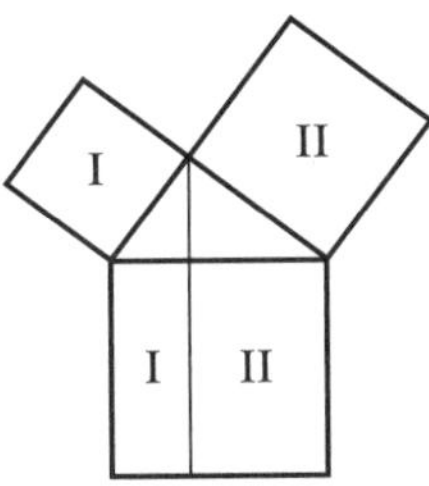

Euclid's bold claim is that the square marked I has the same area as the rectangle marked I, and the square marked II has the same area as the rectangle marked II. If he is correct, then the sum of the areas of the two small squares (I and II) is most definitely the same as the area of the large square (I + II, or III), which is what the Pythagorean Theorem states.

How, though, do we go about demonstrating these two facts? To do this, Euclid draws a second auxiliary line.

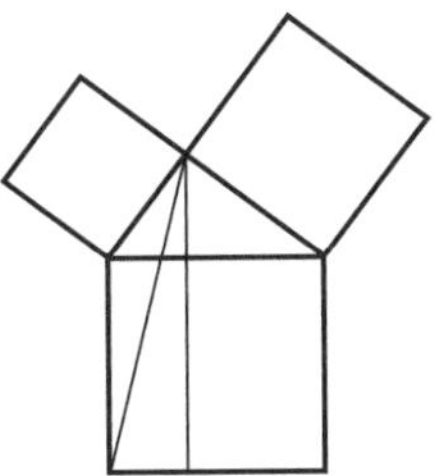

The task at this point is to compare the areas of two different shapes: a triangle and a rectangle. To help see what is going on, we will pull out the relevant part of the shape.

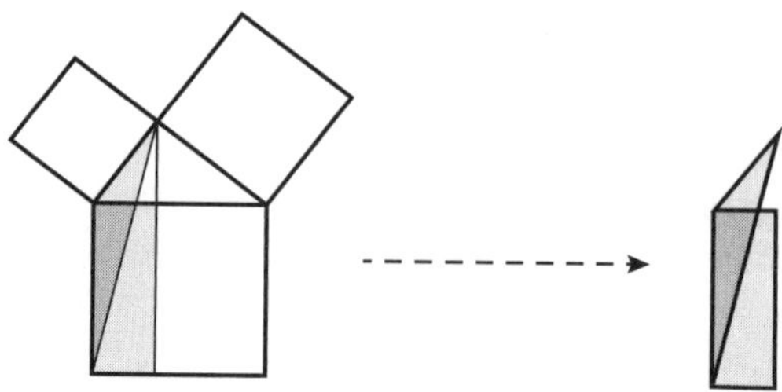

We want to compare the areas of the rectangle and the triangle. To see more clearly, let's tip both of them on their sides and pull them apart. When we do this, we notice something important: the two shapes have the same base and height.

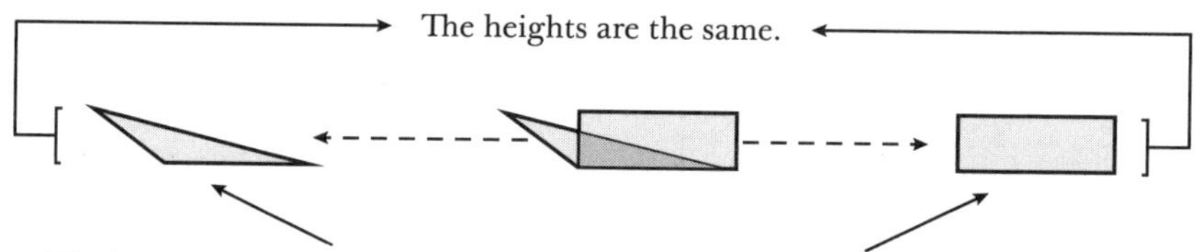

We know from Part I of this book that the area of the triangle is $\frac{1}{2} \times$ base $\times$ height, and that the area of the rectangle is base $\times$ height. Therefore, the area of the rectangle is twice the area of the triangle. This is an important fact to hold in your memory for a bit.

The next step is to draw a different auxiliary line and to pull off two different shapes: a triangle and a square.

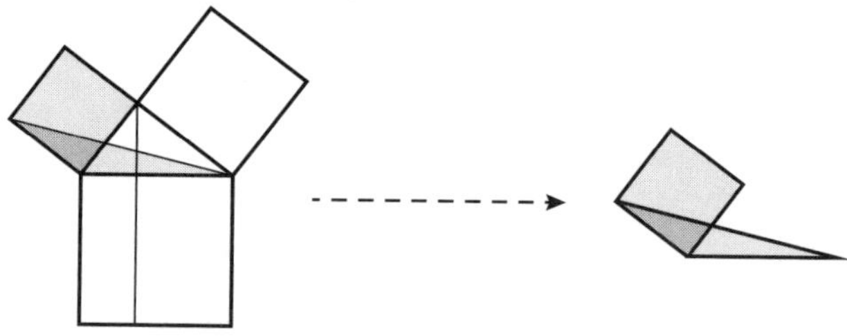

Once again, we are going to tip the shapes and pull them apart, and once again we see that the triangle and the square have the same base and the same height.

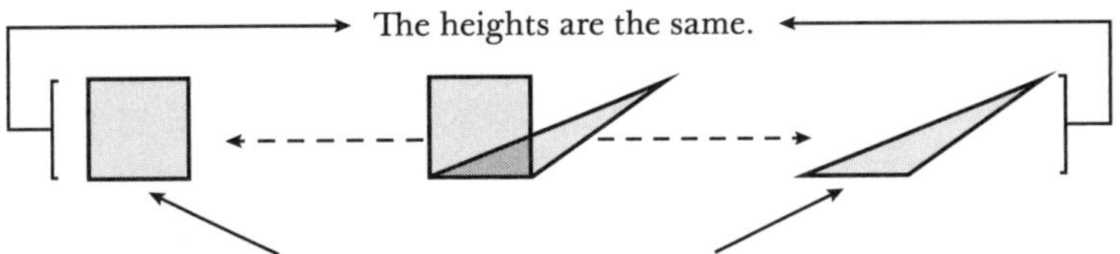

Therefore, the area of the square is twice the area of the triangle.

We now only need to compare the two triangles. This is more difficult, but we can start by drawing both auxiliary lines on the same diagram.

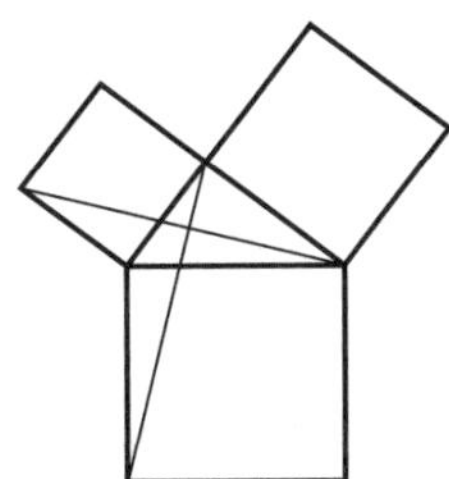

Based on the diagram, we might guess that the two triangles are identical, but convincing ourselves of this is a bit tricky. We start by noticing that there are pairs of sides that are equal because they come from the same square.

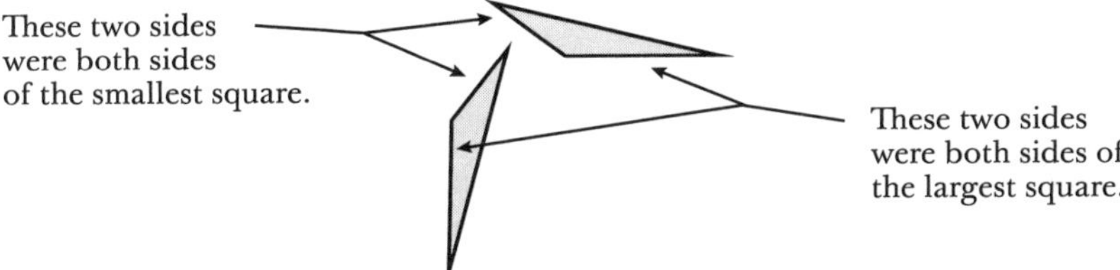

In addition to these two pairs of equal sides, it turns out that the angles formed by them are also equal. To see why, it is helpful to refer to the diagram with both triangles in place. The angles we are talking about are both made up of a right angle and an angle shared by both triangles.

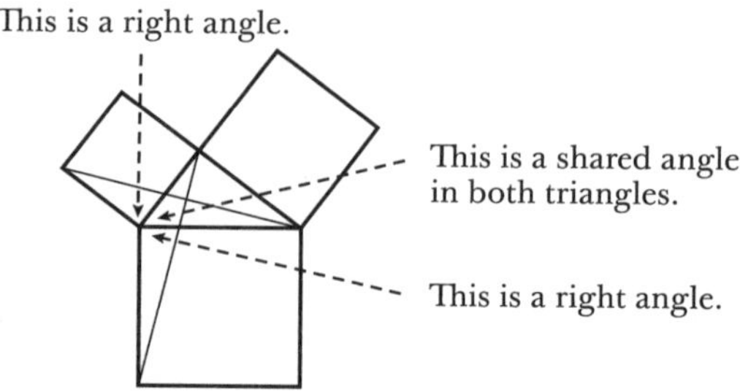

It may be helpful to spin one triangle so that the two are lined up the same way.

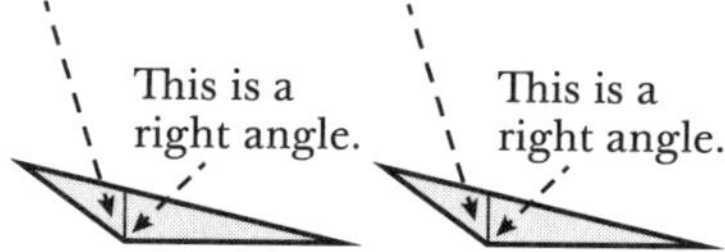

Because the two parts of each angle are equal, the whole of each angle must also be equal. Therefore, we now have two triangles that have a pair of identical angles formed by two pairs of identical sides.

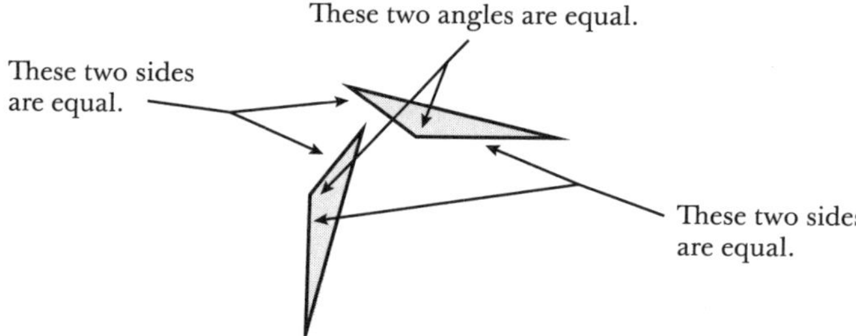

This can only mean that the two triangles are themselves identical. You can think of this in two ways. First, begin by imagining that you draw the first side (maybe the bottom one). Then mark off the angle and draw the second side. If one hundred people did this, their pictures would look identical (assuming perfect drawing skills). We need only connect the two endpoints for the third side of the triangle. In the end, all one hundred triangles will be identical. In other words, when we draw a side, then an angle, and then a side, in that order, *there is only one possible triangle*.

There is another argument that is a bit shorter. We can imagine "sliding" one triangle on top of the other so that the two sides and the angle from each triangle will match up. If we do that, the two triangles will be exactly on top of each other. In other words, there is no way for the third side not to match up as well.[31] Either way we argue it, these two triangles must be identical. If the two triangles are identical, then they have the same area. We are ready to summarize what we know about the rectangle, the square, and the two triangles.

Because both the rectangle and the square are twice the triangle, it can only be that the rectangle and the square are themselves equal. After all of that, remember that these two shapes were the ones marked with a I in the original diagram, so we now know that these two regions have the same area.

Half of our work is done. We need only to show now that the two regions marked II are also equal in area. Fear not, though; we do not intend to repeat all of that work. What Euclid does is to draw a similar pair of auxiliary lines.

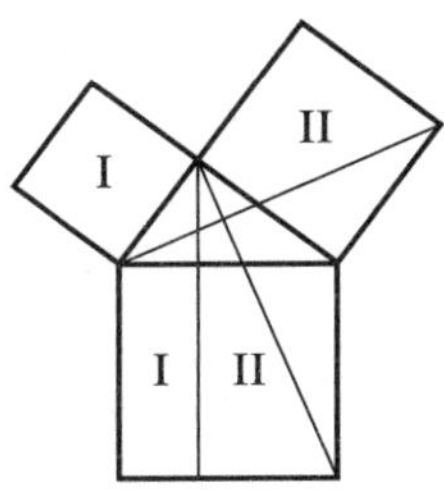

31 This property is Euclid's fourth proposition in Book I of the *Elements* and is more commonly known as the Side-Angle-Side Theorem. Euclid himself gives an argument by "superposition," which is not altogether different from our "sliding" argument. Admittedly, this account is missing some of modern mathematical rigor, but it is sufficient for our purposes here. The goal is an understanding of the proof of the Pythagorean Theorem, so a bit of handwaving on SAS is acceptable.

The work is identical to the first half, but in this case the rectangle and the square are marked II instead of I, and two new triangles are formed. (Can you find them?) Since the argument for this half is identical, it will produce a similar "shape equation" to the one before.

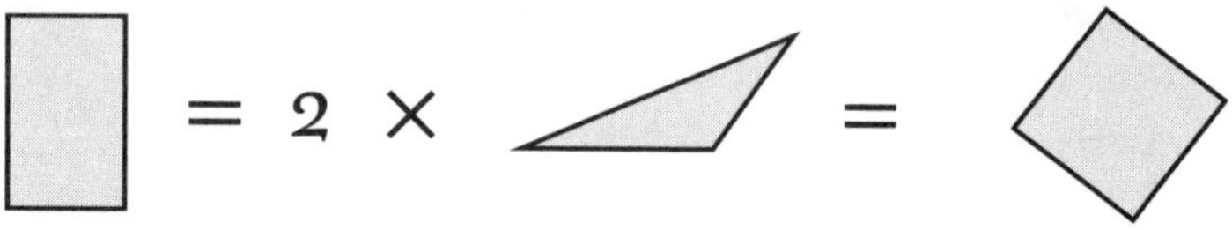

= 2 × =

Once again, the rectangle must be equal in area to the square. Therefore, we now know that the two regions marked II are equal in area. To complete the proof, it is worth repeating why we were interested in regions I and II in the first place. On the one hand, I + II represents the sum of the two smaller squares. On the other hand, I + II represents the area of the largest square, being two rectangles that make it up.

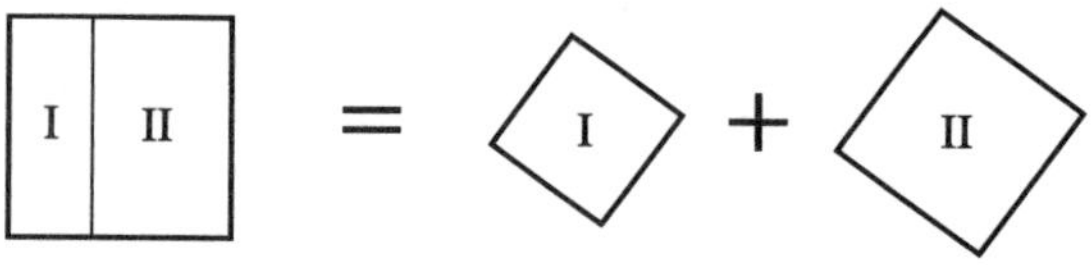

I II = I + II

If the sides of the triangle are a, b, and c, with c being the hypotenuse, then the areas of these three squares are a^2, b^2, and c^2, and we have one of the world's most famous equations: $a^2 + b^2 = c^2$.

There have been hundreds of different proofs of the Pythagorean Theorem. Euclid's, while difficult, is among the most elegant because it requires only geometry and discloses something of the *why* of this remarkable theorem. It shows that the large square is composed of two pieces, each equal in area to one of the two other squares. Euclid's proof, however, is rarely the first one offered to students. That honor normally goes to a proof that is in some ways simpler but does not illuminate the *why* of the theorem as effectively. It begins by organizing four copies of the right triangle to form a square.

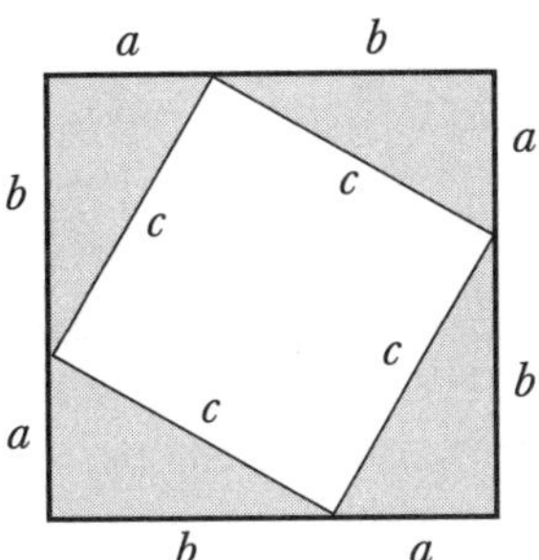

First, we pull apart the shapes and combine the triangles to make two rectangles.[32]

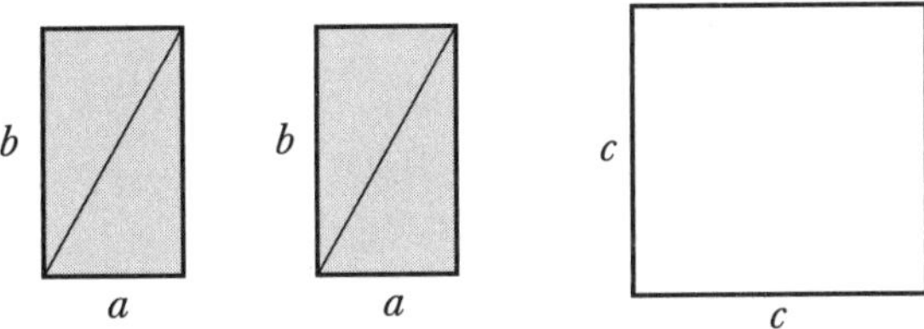

Next, we return to the original diagram and slice it up a bit differently.

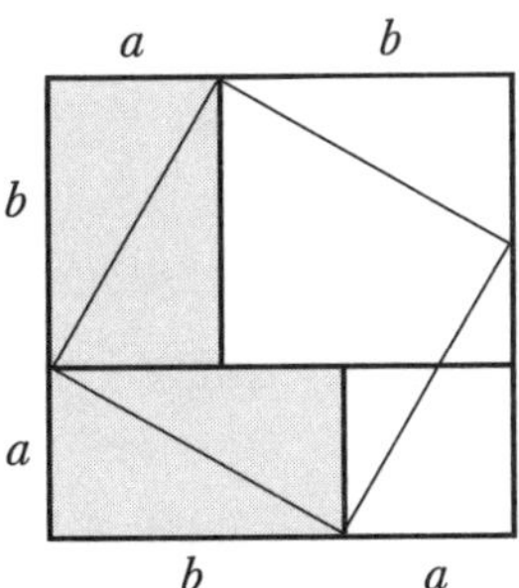

We can see two $a \times b$ rectangles, but we also see two other squares that emerge. We can separate this shape into these component parts.

32 One of the reasons why this proof is not as elegant is that there is some work to be done to show that the diagram is not lying to us. For example, why is the central figure really a square? I leave it to you to work out why those angles must be right angles, but it can be demonstrated using the Triangle Angle Sum Theorem. Moreover, why do the identical right triangles form a rectangle? Again, I leave that to you to justify. It should also be said that this proof is often presented algebraically, starting with $(a+b)^2$ and expanding it, and then using algebraic expressions for the areas of the four triangles and central square. The algebraic version leaves quite a bit to be desired in terms of seeing why the theorem ends up being true. There is a feeling of having received a bit of "luck" with the algebraic terms lining up in just the right way.

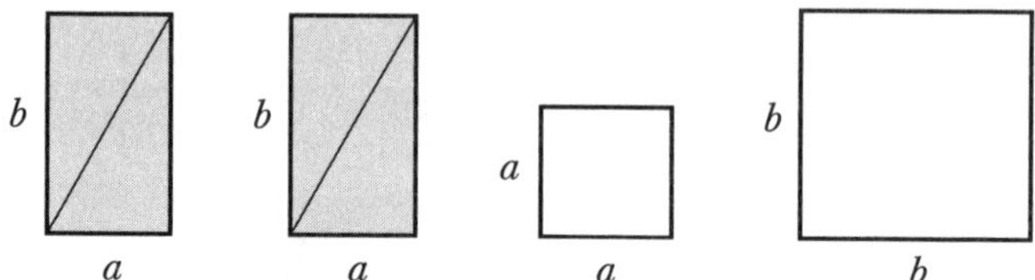

The area of our original shape is the same no matter how we slice it, so the two slicings must be equal to one another.

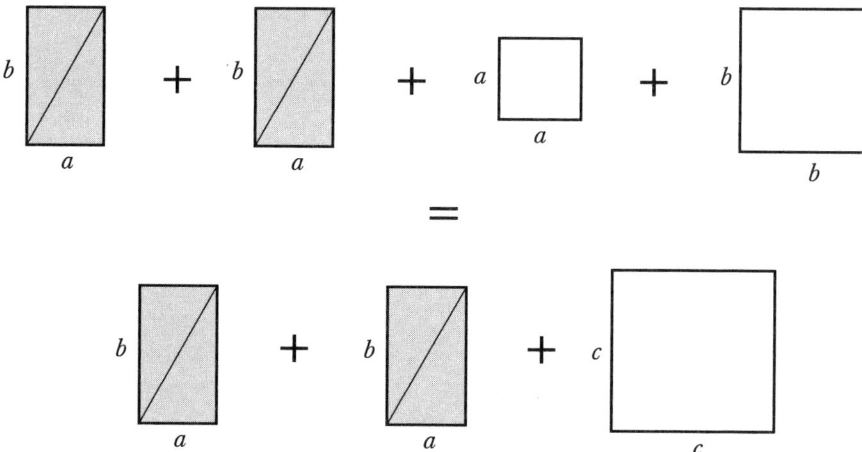

Each side of the equal sign has two identical $a \times b$ rectangles, which means that the other pieces must also be equal.

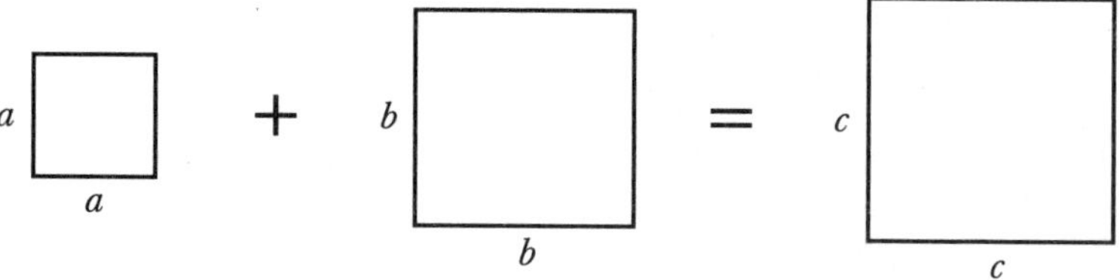

This picture equation of squares is the Pythagorean Theorem. As an algebraic equation it reads $a^2 + b^2 = c^2$, but we should not forget that these variables mean something. They are the lengths of the sides of the triangle, and their squares represent areas of, well, squares!

I would be remiss if I did not mention that a former president of the

United States is credited with his own proof of the Pythagorean Theorem. James A. Garfield published a proof in 1876 in the *New England Journal of Education*. (He was a congressman from Ohio at the time and not yet president.) His proof begins with a similar diagram, but he starts with the formula for the area of a trapezoid and proceeds algebraically.

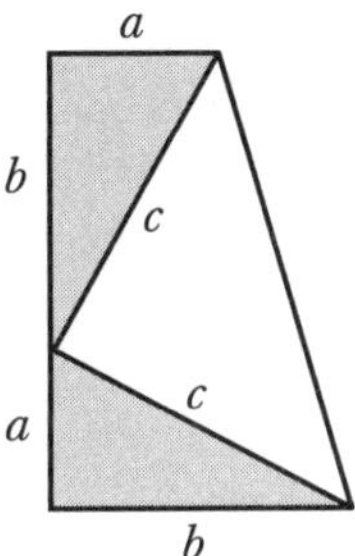

I will let you work out the details, which will look quite similar to our second proof. After all, this trapezoid is half of the square we used.

Finally, while the Pythagorean Theorem, together with its proof, is one of the highest achievements in the *Elements*, it would be an injustice if we did not return to Euclid's generalized theorem referred to by Proclus. In Proposition 31 of Book VI, Euclid considers figures other than squares that are placed on the sides of the triangle.[33]

> *The Generalized Pythagorean Theorem*
> Whenever similar figures are placed on the sides of a right triangle, the area of the largest figure is equal to the sum of the areas of the other two figures.

By *similar* I mean that the figures are all "scale models" of each other. While many people are familiar with the Pythagorean Theorem, this more powerful property is not nearly as well known. If the original theorem is surprising, its generalization is more so. Euclid says that we can build *any figure we want* on one side of the triangle and then build figures on the other two sides that are scale models of the original. The result is that the Pythagorean relationship about areas still holds. Euclid only built

33 The actual proposition reads, "In right-angled triangles, the figure on the side subtending the right angle is equal to the similar and similarly described figures on the sides containing the right angle."

polygons onto the triangle, but we can even use shapes with curves.

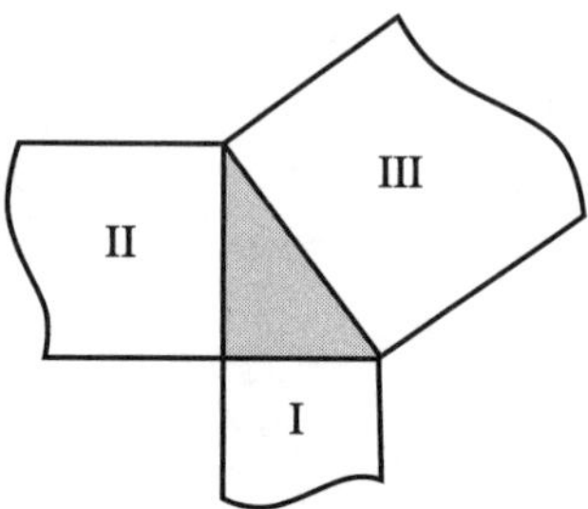

It turns out that the area of the larger shape is equal to the sum of the other two areas, *no matter the shape*, i.e., I + II = III. The Pythagorean Theorem is a special case of this proposition with the shapes being squares.

CHAPTER 4
THE EXISTENCE OF IRRATIONAL NUMBERS

GOING ALL the way back to our grammar school mathematics classes, we have performed countless calculations using just four basic operations: addition, subtraction, multiplication, and division. Along the way, we probably have taken something for granted: the result of these four operations is almost always another number. In other words, if we take two numbers and add them together, the result is some number that actually exists. The same thing holds for subtraction, multiplication, and division—mostly. We say "mostly" because of two caveats.

First, results such as $5 \div 0$ do not exist. This has to do with division being the opposite of multiplication. For example, if we are looking for the result of $6 \div 2$, we ask ourselves, "What is a number that when multiplied by 2 gives us 6?" The answer is 3 because $3 \times 2 = 6$. When calculating $5 \div 0$, we ask, "What is a number that when multiplied by 0 gives us 5?" The answer this time is that no such number exists because anything multiplied by 0 is 0 and therefore can never be 5. In mathematics, we say that $5 \div 0$ is *undefined*. The case of $0 \div 0$ is uniquely puzzling. What is a number that when multiplied by 0 gives us 0? The answer is that we can use any number we want. After all, 0 times anything is 0. This is why mathematicians call the result of $0 \div 0$ *indeterminant* rather than *undefined*. This distinction turns out to be very important in calculus.

The second caveat concerns subtraction, at least in elementary school. If the number being subtracted is greater than the original number, we have a problem. Without the concept of negative numbers, the result of $3 - 7$ is nonsense. In a very real way, the idea of negative numbers is a mathematical construct created to deal with the fact that $3 - 7$ is not actually a number otherwise known. Negative numbers are quite familiar to us, and yet it is difficult to think about the existence of these numbers in the "real world." We might think that something like debt can be thought of as a negative number. Even then, when I am \$100 in debt, I don't actually have −\$100. Instead, I have \$0, and I owe someone else \$100. This is not to deny the existence or even usefulness of negative numbers. It is, however, to highlight both the beauty of mathematical creations and the remarkable ability of mathematics to deal with things that are purely mental objects, which are no less real than material objects. This will also be an important fact later when we consider the complex numbers, which are no less "real" than negative numbers.

With those two caveats out of the way, we return to the interesting assumption we all make: the result of any number added to, subtracted from, multiplied by, or divided by any other number is itself *another number*. Mathematicians use the word *closed* to talk about this relationship of numbers to operations. The whole numbers are closed under addition and multiplication because the sum or product of two whole numbers is another whole number.[34] But we can also ask something of a backwards version of the closure question: "Can all numbers be built up by starting with only 0 and 1 using only our four basic operations?"

In other words, if I start with only the number 1, I can use addition to get 2, 3, 4, 5, etc. For example, 2 is $1+1$, 3 is $2+1$, etc. I can use subtraction to get 0 and the negative numbers -1, -2, -3, etc. For example, 0 is $1-1$, -1 is $0-1$, and -2 is $0-2$. Putting the positive and negative numbers together, we get the set of integers (..., -2, -1, 0, 1, 2, 3, ...). We did this only with addition and subtraction, and it turns out that multiplication does not get us any new numbers. Any integer times any other integer is still an integer, i.e., the integers are closed under multiplication. Division, however, is a wholly different story, pun intended. The result of $6 \div 2$ is indeed an integer, but the result of $2 \div 6$ is *not*. Rather, it is a new number altogether: $\frac{1}{3}$. With division, we now have *a lot* more numbers. The question is, with division in place, do we now get *all* the numbers?

The results of division problems are fractions, or what mathematicians call the *rational numbers*. (The term "rational" comes from *ratio*, a Latin word meaning "an account" or "a reasoning.") With these "new" numbers in place, the natural question is "Do we have them all?" In other words, are all numbers expressible as fractions?[35] The Greeks were very concerned about this question. Many postulated that the answer was yes—that all numbers could be written as a quotient of whole numbers. Said differently, they thought the rational numbers make up all the numbers that exist. It was quite a surprise when they found out that this is not the case. The fly in

34 We have to be precise in using the word "closure." The whole numbers are not closed under subtraction because $3-7$ is not a whole number. But the integers (..., -2, -1, 0, 1, 2, 3, ...) are closed under subtraction. The set of rational numbers is not closed under division because of the problem of dividing by zero. But "the set of all non-zero rational numbers" is closed under division.

35 Note that integers, such as 6, are expressible as fractions: $6=6\div 1$, or $\frac{6}{1}$. Even 0 is $0\div 1$, or $\frac{0}{1}$.

the ointment was the diagonal in a right triangle with legs both of length 1.

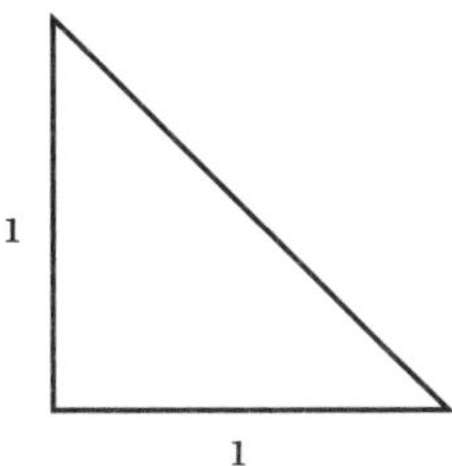

We should pause here to note how "reasonable" this number is. For the Greeks, a "real" number (not to be confused with the modern terminology "real number") is one that could be constructed geometrically. Numbers and shapes were not really distinct things for the Greeks: the number 6 really meant a segment of length 6. They used circles (the tool for which was a compass) and lines (the tool for which was a straightedge) to draw certain geometric objects, and they had known for a long time how to construct segments whose lengths were the result of adding, subtracting, multiplying, or dividing the lengths of two other segments.

But they also knew that a segment could be copied and placed at a right angle to another segment, and that the endpoints of the segment could be joined. Therefore, the above triangle was a very real object for them, which means that the length of the diagonal was a very real number. In fact, for the Greeks, this number would have been far more "real" than -4. We cannot, after all, construct a segment of length "-4 units." We *can* construct a segment that is the diagonal of a right triangle if the length of each leg is 1.

The looming question became "Is the length of the diagonal in the triangle above the result of a division problem?" To answer this question, we rely first on the Pythagorean Theorem. If we call the diagonal (or the hypotenuse) c, then the Pythagorean Theorem tells us that $1^2+1^2=c^2$. In other words, $c^2=2$. As we explore the implications of this, buckle up. This is the first proof that has some algebra in it, but fear not: we will take it slowly.

In ancient Greece, the working presumption was that c was indeed a fraction, which means it can be written as $c=\frac{M}{N}$, where M and N are whole numbers. The first thing to do is to reduce the fraction $\frac{M}{N}$ to lowest terms, which we will call $\frac{m}{n}$. Therefore, $c=\frac{m}{n}$, and m and n have no factors in common. Because $c^2=2$, it must be that,

$$\left(\frac{m}{n}\right)^2 = 2\,.$$

The left side of the equation is the result of squaring or multiplying a number by itself.

$$\left(\frac{m}{n}\right)^2 = \frac{m}{n} \times \frac{m}{n} = \frac{m \times m}{n \times n} = \frac{m^2}{n^2}$$

Therefore,

$$\frac{m^2}{n^2} = 2\,.$$

This means that 2 is the result of dividing m^2 by n^2, which is another way of saying that "2 times n^2 is equal to m^2," or

$$m^2 = 2n^2.$$

I now make a bold claim: the number m^2 is even. Why is that? What does it mean to be an even number? It means that the number is divisible by 2. Equivalently, it means that we can write the number as "2 times another whole number" or that 2 is a factor of the number. Because m^2 is "2 times n^2," it must be that 2 is a factor of m^2, and therefore m^2 is an even number. I now make a second bold claim: if m^2 is even, then m itself must be even. This is because an even number multiplied by an even number is always even, and an odd number multiplied by an odd number is always odd. Therefore, m and m^2 must be either both even or both odd. Because m^2 is even, it follows that m must be even. This means that m can be written as "two times another number." In other words, m can be written as "$2k$," where k is a whole number. But if $m = 2k$, then $m^2 = m \times m = 2k \times 2k$, which is $4k^2$.

This is a big moment. We know that $m^2 = 2n^2$, and we know that $m^2 = 4k^2$. Therefore,

$$4k^2 = 2n^2.$$

We can divide both sides of the equation by 2 to find that

$$2k^2 = n^2.$$

This tells us that n^2 is an even number by the very same rationale that told us that m^2 was an even number. Further, because n^2 is an even number, n must also be an even number, again by the same reasoning that we used to determine that m was an even number.

We now have the curious fact that both m and n are even numbers. It is curious because it in fact cannot be true. This is because the fraction $\frac{m}{n}$ was in lowest terms, which means that m and n cannot have any factors in common. If m and n were both even, then they would have a factor of 2 in common.

We are forced to conclude, then, that the number c cannot be written as a fraction of whole numbers. If it could, then that fraction would simultaneously be both in lowest terms and not in lowest terms, which is a logical contradiction. Because of this, we have discovered the existence of the *irrational numbers*. Or at least we have discovered the existence of *one* irrational number, namely $\sqrt{2}$.

> *The Existence of Irrational Numbers Theorem*
> There are numbers that exist that are not rational.
> That is, there are numbers that cannot be written as
> a fraction of integers. One such number is $\sqrt{2}$.

It turns out that the square roots of most numbers are irrational. In fact, as long as the number is not a perfect square (1, 4, 9, 16, 25, etc.), then its square root is irrational. For example, $\sqrt{3}$, $\sqrt{5}$, $\sqrt{6}$ and $\sqrt{7}$ are all irrational. ($\sqrt{4}$ is not because $\sqrt{4}=2$.) These are not the only irrational numbers; we will discuss later the fact that the number π is irrational. For now, however, we leave you with one more irrational number. The solution to the following equation is irrational.

$$2^x=3$$

The argument here is actually a little simpler than that used for $\sqrt{2}$, but it does require some algebra of exponents.[36] Again, to show that x cannot

36 Admittedly, the difficult task in this case is convincing ourselves, even informally, that a real number solution to this equation actually exists at all. We convinced ourselves that $\sqrt{2}$ was "real" by sketching a right triangle and identifying this number as the length of the hypotenuse. It turns out that the number x under discussion is not actually constructible, which means a segment of this length *cannot* be drawn with a compass and a straightedge. Therefore, and I am speculating, the Greeks may not have even considered x a number at all, whereas $\sqrt{2}$ was perfectly reasonable. Nevertheless, the solution x is a real number, and I thought the proof was worth including.

be rational, we start by assuming that it is. Therefore, write x as a fraction of whole numbers: $x = \frac{m}{n}$. Therefore,

$$2^{m/n} = 3.$$

Raise both sides to the nth power and simplify.

$$\left(2^{m/n}\right)^n = 3^n$$

$$2^m = 3^n$$

The problem is that the left side is even because a string of 2s multiplied together will always be even, and the right side is odd because a string of 3s multiplied together will always be odd. An even number can never equal an odd number. Therefore, our starting assumption cannot be correct because it leads to this absurdity.[37] We can therefore conclude that x cannot be a rational number.

Our proof for the irrationality of $\sqrt{2}$ is in Book X of Euclid's *Elements*, and Aristotle mentions it in Book I of his *Prior Analytics*. But it is often Hippasus, a Pythagorean, who is credited with the first discovery of this result. That discovery is doubted by some historians, as is the legend that Hippasus was intentionally drowned at sea for his accomplishment as both a punishment for violating the deeply held mathematical-philosophical beliefs of the Pythagoreans and an attempt to keep the result hidden. Whether or not the legend is true, it is certain that the discovery of perfectly reasonable irrational numbers shook the foundations of how mathematicians thought about the very nature of numbers.

37 Okay, it turns out there is a solution, which is $m = n = 0$, because $3^0 = 2^0 = 1$. But remember that m and n represented the numerator and denominator of x, so we cannot have $n = 0$.

CHAPTER 5
IRRATIONAL DECIMALS

SOMETIME IN elementary school, students learn how to turn fractions into decimals. The motivation for learning this seems to have something to do with decimals being "nicer to work with." This is too bad, actually, for fractions themselves are beautiful things and at least in some ways are more natural to work with than decimals. This statement should take nothing away from what a decimal representation has to offer, but it should encourage us to think about what the interesting questions about decimals might be. In other words, both fractions and decimals are fabulous and interesting and deserve to be investigated, but the *reason* for studying one or the other should not be that one is "easier to work with."

The mere fact that the same number can have very different representations is a beautiful thing. The fact that a number like $\frac{3}{4}$ can be written as the sum of other fractions with powers of 10 in the denominator, $\frac{7}{10}+\frac{5}{100}$ or 0.75, is pretty neat. It is not at all clear that all fractions can be written this way. We are perhaps too familiar with elementary school math to find this surprising, but it is worth pausing to think about why it is the case. Let's pretend for a moment that we don't already know that $\frac{3}{4}$ is the same as 0.75. How were we taught to convert this fraction to a decimal way back when? The process is first to see a fraction as a division problem; that is, $\frac{3}{4}$ is really the same as $3 \div 4$.

Wait. What? A fraction is a number that is the result of a division problem? In elementary school, the first time we encounter $\frac{3}{4}$ we define it something like this: "Take a picture, maybe a circle, and split it into four equal parts, maybe pie slices. Then shade three of them. This picture represents $\frac{3}{4}$." This is a rough definition, but it is not too far off from how we explain fractions to young students for the first time. The definition is something like "parts of a whole." This is not mere child's play; it is fairly close to how the Greeks understood fractions. They are not so much numbers as *relationships between numbers*. They are *ratios*. Remember in the introduction how the Pythagoreans understood music as the branch of mathematics that studies quantity in relationship? This is because musical intervals are really about the ratios of frequencies. Only with modern technology can we actually define a note such as A as 440 Hz. For the ancients, such isolation of pitch would not have made sense; there really was no "A." What *did* exist was a "perfect fourth," or pitches (plural) in the ratio of $\frac{3}{4}$.[38] It was the physical objects—the lengths of two strings or the weights of two Pythagorean hammers—that formed this ratio. We now know that it is the lengths of the sound waves for the two notes, or inversely their frequencies, that are involved. For the Greeks, this wasn't just about music. All fractions (rational numbers) first and foremost describe a relationship between two numbers.[39]

Somewhere along the line we switch from telling students that $\frac{3}{4}$ means "one whole, split into four, with three parts selected," to "*three* wholes, split into four, with *one* part selected," which is the same as $3 \div 4$. It turns out these two things are the same, but this is not immediately evident.

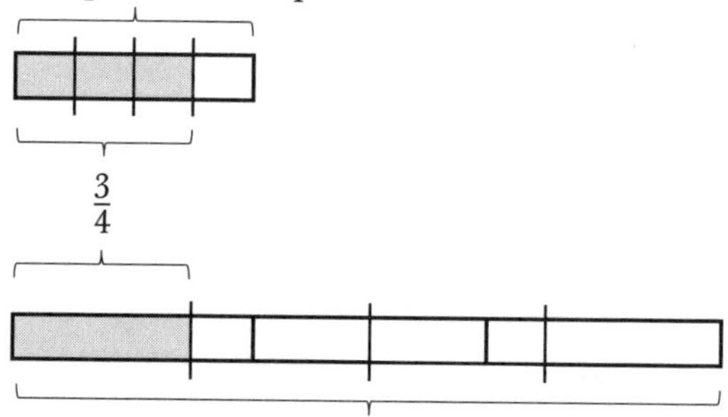

We can *see* that this equality is true for $\frac{3}{4}$ in the picture, but does this work for all numbers? Is it an amazing fact of fractions that "one whole, split into *b* parts, with *a* parts selected" and "*a* wholes, split into *b* parts, with one part selected" are always the same! We will not go through an argument for this, but I encourage you to think through it on your own. For our purposes, it is enough to note the fact.

The second conceptualization is the one that allows us to think about fractions as *numbers*. "Three wholes split into four parts" is the definition of division: $3 \div 4$. Once we have fractions as numbers, then we can think about performing operations on them (addition, subtraction, multiplication, division, etc.), comparing them, and, most relevant to our current

38 The perfect fourth can either be described as $\frac{3}{4}$ or $\frac{4}{3}$, depending on which string/hammer/frequency comes first.

39 This is an oversimplification, of course, of the Greek understanding, but it is sufficient for our purposes.

conversation, converting them to decimals.[40]

Now that we have settled the fact that $\frac{3}{4}$ really *is* $3 \div 4$, we can proceed to convert it to a decimal using that elegantly efficient algorithm of long division.

$$\begin{array}{r} 0.75 \\ 4\overline{)3.00} \\ -2\,8 \\ \hline 20 \\ -20 \\ \hline 0 \end{array}$$

The curious thing is that this tidy division sometimes takes a bit longer. In fact, sometimes it simply never ends. Consider the fraction $\frac{41}{333}$.

$$\begin{array}{r} 0.1231 \\ 333\overline{)41.0000} \\ -333 \\ \hline 770 \\ -666 \\ \hline 1040 \\ -999 \\ \hline 410 \\ -333 \\ \hline 77 \end{array}$$

We see the same number, 410, come up after a few subtractions. At this point, we know that the process will repeat without end. After 410, the results of the other subtractions (and dropping down of the 0) are 770, 1040, and then 410 again. The decimals on top will also continue to repeat, so representation for $\frac{41}{333}$ is 0.123123123123…, which we usually write as $0.\overline{123}$.

At this point, we know that some fractions like $\frac{3}{4}$ will have terminating decimals, and some fractions like $\frac{41}{233}$ will have repeating decimals.[41] Is there any other possibility? Maybe we can have decimals that neither repeat *nor* terminate. While certainly an interesting question, it turns out

40 The mere fact that we do things like add fractions together should be mystifying if we are used to thinking of fractions only as "parts of a whole." If $\frac{3}{4}$ is a concept represented by a certain picture, and $\frac{2}{5}$ is another concept represented by a different picture, what business have we of adding two concepts together, and what does the result even mean? This mystifying and marvelous transition from "parts of a whole" to "fractions as numbers" is likely one reason why fractions are so difficult for young students.

41 One of my favorite problems to ask people to think about is to try and predict whether a fraction will repeat or terminate just based on its numerator and denominator. A harder task is how to tell the length of the header, or the nonrepeating decimals up front. For example, there are three decimals that come before the repetition starts in $0.613\overline{45}$, i.e., 6, 1, and 3. A second hard task is how to tell the length of the period of repetition, i.e., the fact that there are two digits that repeat in this decimal, 4 and 5.

not to be possible for fractions (rational numbers). To see why, it is easier to consider a fraction like $\frac{2}{7}$.

At each step of the long division algorithm, we are asking, "How many times does 7 go into ___?" In other words, we are doing little "division by 7" steps. When we subtract at each step, we are finding the *remainder* when that number is divided by 7.

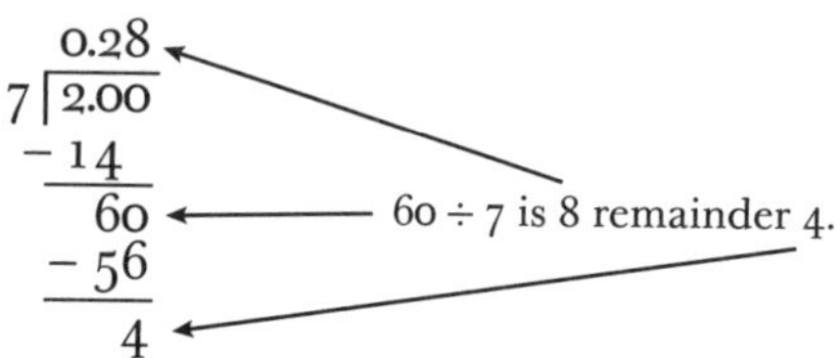

The key thing to notice is that when we divide by 7, there are only seven possible remainders: 0, 1, 2, 3, 4, 5, or 6. At each step after the subtraction (before dropping down the 0), we can only ever get one of these seven numbers. If we get 0, then the decimal will terminate. If we never get 0, then *eventually a remainder has to reappear*. With only seven possibilities, at some point the same number has to come up. When that particular remainder reappears, the "divide by 7" exercise starts repeating. For example, in the problem above, the first remainder was 6. At that point, once the zero gets dropped down, we are figuring out 60 ÷ 7. If a 6 ever shows up again as a remainder, we will again be figuring out 60 ÷ 7, causing the whole thing to repeat.[42] With only seven possible remainders, eventually at least one of the remainders must repeat, and when that happens, the digits will also repeat.

We now have a nice theorem about fractions (rational numbers).

> *Decimal Expansions of Rational Numbers Theorem—Part 1*
> If a rational number is converted to a decimal,
> the digits will either repeat or terminate.

We can also go backwards. A terminating decimal such as 0.6789 can be easily turned into a fraction by either writing it out literally according to place value, $\frac{6}{10}+\frac{7}{100}+\frac{8}{1{,}000}+\frac{9}{10{,}000}$, or by taking the shortcut and

42 Note that the 6 itself does not have to repeat, though in this case it does. There are rational numbers that have decimal expansions such as 0.6543434343..., where the first two digits do not show up ever again. The point is not that the 6 has to repeat but that *some remainder* has to repeat eventually because there are only seven possibilities.

writing it as $\frac{6,789}{10,000}$. This fraction may not be written in simplest form, but it does show that the decimal can be written as a fraction. This is fine for decimals that terminate, but what about converting repeating decimals to a fraction? What if we start with a number such as 0.33333333…? This repeating decimal is very familiar, but let's pretend that we do not yet know its fraction equivalent. Converting repeating decimals to their equivalent fractions is one of the most ingenious ideas that used to be a standard part of elementary and middle school mathematics curricula. Occasionally it still shows up in a textbook, but more and more it is being left out, which is really a shame.

The idea is to set up a subtraction problem.

$$\begin{array}{r} 3.3333333\cdots \\ -\ 0.3333333\cdots \\ \hline 3.0000000\ldots \end{array}$$

Each of the infinite number of decimal place values subtracts to get 0. The real beauty is found in the relationship between the first number (3.3333333…) and the number being subtracted (0.3333333…). They are only off by exactly one decimal place, or a factor of ten. In other words, if $\blacksquare = 0.3333333\ldots$, then $10 \times \blacksquare = 3.3333333\ldots$. This, of course, only works because there are an infinite number of decimal places. We can rewrite our subtraction problem.

$$10 \times \blacksquare - \blacksquare = 3$$

Ten times any number minus that number is always nine times that number. For example, $10 \times 7 - 7$ is 9×7, which is just one fewer copy of 7. Therefore, $10 \times \blacksquare - \blacksquare = 9 \times \blacksquare$, so our subtraction problem can be simplified.

$$9 \times \blacksquare = 3$$

We need only divide both sides by 9 to find that $\blacksquare = \frac{3}{9}$ or $\frac{1}{3}$.[43]

43 We do need to be a bit careful with arithmetic involving an infinite number of terms. There is a well-known "false proof" that starts with …999999, which in this case has an infinite number of 9s and no decimal place. Multiplying this number by 10 gives us …9999990. If we go through an argument similar to the one for 0.333333…, we will end up with …999999 being equal to −1. But this is because the number …9999999 is actually infinity. Mathematicians would say that it is a series that *diverges*. The series for 0.333333… *converges*, and because of this we can do the arithmetic that is outlined here. The proof for its convergence is beyond the scope of this book.

But ■ was precisely the infinitely repeating decimal we were looking for: 0.3333333….

$$0.3333333\cdots = \frac{1}{3}$$

We can use this technique with any number. While we will not go through the details, it is worth thinking through how to go about converting decimals such as 0.345345345345…. For this, we set up a different subtraction problem.

$$\begin{array}{r} 345.345345345\cdots \\ -\quad 0.345345345\cdots \\ \hline 345.000000000\ldots \end{array}$$

The first number, 345.345345345…, is 1,000 times the number being subtracted, 0.345345345…. I leave it to you to finish this problem.

How do we deal with decimals that have a few extra digits in a header before the repetition starts: 0.678989898989…? The computation is a bit trickier, but the same ideas work. We first need a subtraction problem that eliminates all of the decimal places by taking advantage of the repetition.

$$\begin{array}{r} 6{,}789.898989\ldots \\ -\quad 67.898989\ldots \\ \hline 6{,}722.000000\ldots \end{array}$$

If ■ is our original number (0.678989898989…), then this subtraction problem is

$$10{,}000 \times ■ - 100 \times ■ = 6{,}722$$

or

$$9{,}900 \times ■ = 6{,}722.$$

If we divide both sides by 9,900, we see that $■ = \frac{6{,}722}{9{,}900}$ or $\frac{3{,}361}{4{,}950}$.

Therefore, $0.678989898989\ \ldots = \frac{3{,}361}{4{,}950}$.

We can use this technique to turn any repeating decimal, even those with headers, into fractions. It is precisely *because* the decimals are infinitely repeating that we can move the decimal point to create two different numbers with identical decimal places. When we subtract these two values, all of the decimal places go away. Because this process can always be completed, we have second part to our theorem about decimal expansions.

> *Decimal Expansions of Rational Numbers Theorem—Part 2*
> If a number has decimal digits that repeat or terminate, then the number is a rational number.

The two parts of the theorem are *converses* of one another. The converse of "if p then q" is "if q then p." Sometimes a theorem can be true, but its converse false. For example, "If a shape is a square, then it is a rectangle" is true because all squares are also rectangles. But the converse, "If a shape is a rectangle, then it is a square," is false because some rectangles are not squares. Other times, a theorem and its converse are both true.[44] When this happens, the combination is particularly powerful. In this case, we now know that rational numbers *and only* rational numbers have decimals that either repeat or terminate. That allows us to make a very bold claim about *irrational* numbers.

> *Decimal Expansions of Irrational Numbers Theorem*
> The irrational numbers are numbers whose decimal expansions have digits that neither repeat nor terminate.

This theorem is astounding! If we convert $\sqrt{2}$ to a decimal (1.41421...) and somehow start calculating its digits, the process will never stop, *and we will never get repetition*. If the decimals stopped, we could write a simple fraction for it, making $\sqrt{2}$ rational, which we know not to be the case. If the decimals repeated, we could go through the above process to write it as a fraction, again causing it to be rational.

Often "a decimal that does not repeat and does not terminate" is offered

44 We saw in an earlier footnote that both the Pythagorean Theorem and its converse are true. These make up Propositions 47 and 48 at the end of Euclid's *Elements*.

as a *definition* of an irrational number. This is problematic for two reasons. First, it ignores the nomenclature *irrational*, meaning "not rational." A rational number was defined as one that can be written as a fraction of integers, so irrational must mean a number that *cannot* be written as a fraction of integers. The definition should have nothing to do with decimal places but rather with fractions. Second, and more importantly, it robs us of the shocking result that is the theorem.

Note that the theorem does not mean that the decimals of an irrational number are *without pattern*. The digits of $\sqrt{2}$ seem to be fairly random, but there are some irrational numbers that have very clear patterns.

$$0.101001000100001000001000000 1\ldots$$

$$0.123456789101112131415161718192021222324252627\ldots$$

The first number has a 1 followed by one 0, then a 1 followed by two 0s, then a 1 followed by three 0s, and so on. The second number is simply counting 1, 2, 3, 4, 5, 6, 7, 8, 9, 10, 11, 12, etc., in its digits. How do we know that these numbers are not rational? This is the power of our theorem. The digits do not stop, and they do not repeat. Therefore, the number *must* be irrational. Nevertheless, there are clear patterns to the digits.

CHAPTER 6
$0.\overline{9}$

IN THE previous chapter, we talked about how to convert repeating decimals to fractions, and we saw that 0.3333333…, or $0.\overline{3}$, is equal to $\frac{1}{3}$. On the one hand, this is not surprising, and for no fewer than three reasons. First, it is a familiar fact, one we have likely never questioned. Second, we gave a convincing argument for it during our exploration of repeating decimals. Third, when in doubt we could divide 1 ÷ 3 to see that the decimals keep coming up as 3. Thinking about this from another perspective, however, there *is* something surprising going on. Let's write out the expansion in terms of its place values.

$$0.3333333\cdots = \frac{3}{10} + \frac{3}{100} + \frac{3}{1{,}000} + \frac{3}{10{,}000} + \frac{3}{100{,}000} + \frac{3}{1{,}000{,}000} + \frac{3}{10{,}000{,}000} + \cdots$$

What we have is an infinite number of things being added together—*and the sum is equal to* $\frac{1}{3}$. How can an infinite number of things be added together, and the sum be *finite*? This is one of the marvels of infinity and is the basis of much of calculus. To be sure, there are certainly cases where an infinite number of terms are added together and the result is *not* finite. For example, 1+1+1+1+… is most certainly infinite. In the sum for 0.3333333…, though, the key is that each term is getting smaller and smaller. In fact, each term is getting closer and closer to 0 or "infinitely small." That is the *only* way that this has hopes of working—if the sum of an infinite number of things is to have a finite result, then each term must get progressively and infinitely small.

Such a thing happens with other sums, not just decimal expansions. For example, consider a sum in which the denominators keep doubling so that the terms become infinitely small.

$$\frac{1}{2} + \frac{1}{4} + \frac{1}{8} + \frac{1}{16} + \frac{1}{32} + \cdots$$

We can use a technique to work out this sum that is similar to the technique we used for repeating decimals. Let's start with the following subtraction problem.

$$\begin{array}{r} 1+\frac{1}{2}+\frac{1}{4}+\frac{1}{8}+\frac{1}{16}+\frac{1}{32}+\cdots \\ -\quad \frac{1}{2}+\frac{1}{4}+\frac{1}{8}+\frac{1}{16}+\frac{1}{32}+\cdots \\ \hline 1 \qquad\qquad\qquad\qquad\qquad\qquad \end{array}$$

If ■ is our original number $(\frac{1}{2}+\frac{1}{4}+\frac{1}{8}+\frac{1}{16}+\frac{1}{32}+\cdots)$, then this subtraction problem is

$$2\times\blacksquare-\blacksquare=1,$$

or

$$\blacksquare=1.$$

This must mean that $\frac{1}{2}+\frac{1}{4}+\frac{1}{8}+\frac{1}{16}+\frac{1}{32}+\cdots=1$.

Geometrically, the sum looks like a 1×1 square that keeps getting "halves" filled in.

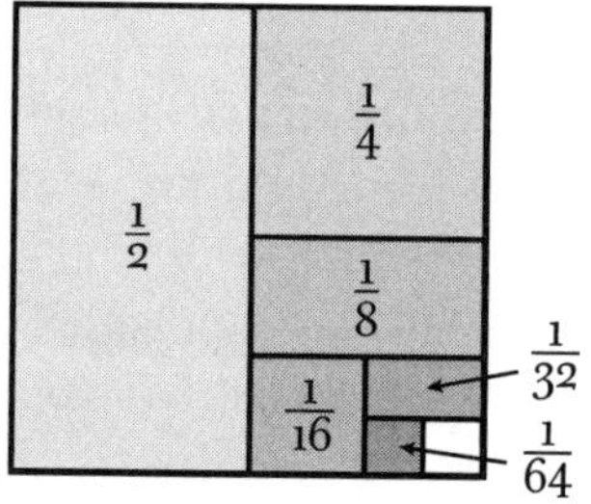

True, there is still a corner that is not filled in, but I have only shown six terms. We will keep cutting the remaining piece in half over and over again. With all of the infinite number of terms added, the entire 1×1 square will be completely filled in. When people see this picture, they are bothered and often say, "I understand that it gets closer and closer to 1, but there is always a little piece missing." This is reasonable, but it is also "finite" thinking. It is true that if we stop the process at any point, the square will not have been filled in, and the sum will be less than 1. But if we continue it with an *infinite* number of terms, the square fills in completely.

As with $\frac{3}{10}+\frac{3}{100}+\frac{3}{1,000}+\frac{3}{10,000}+\frac{3}{100,000}+\cdots$, the sum $\frac{1}{2}+\frac{1}{4}+\frac{1}{8}+\frac{1}{16}+\frac{1}{32}+\ldots$ only has the possibility of being finite because the individual terms become infinitely small. But that precondition alone is not enough. There are examples of an infinite number of terms becoming infinitely small, and yet when they are added together the sum is infinite. The most famous example of this kind of sum is called the Harmonic Series.

$$1+\frac{1}{2}+\frac{1}{3}+\frac{1}{4}+\frac{1}{5}+\frac{1}{6}+\cdots$$

The individual terms in this sum become infinitely small, but it turns out that the sum is still infinite! It's amazing, especially if you start actually performing the additions. The sum grows so slowly that it seems hard to believe it will ever get above 10 let alone head off towards infinity.[45] To convince ourselves that it can indeed get as large as we want, we can group the terms strategically.

$$1+\frac{1}{2}+\left(\frac{1}{3}+\frac{1}{4}\right)+\left(\frac{1}{5}+\frac{1}{6}+\frac{1}{7}+\frac{1}{8}\right)+\left(\frac{1}{9}+\frac{1}{10}+\frac{1}{11}+\frac{1}{12}+\frac{1}{13}+\frac{1}{14}+\frac{1}{15}+\frac{1}{16}\right)+\left(\frac{1}{17}+\frac{1}{18}+\cdots+\frac{1}{32}\right)\cdots$$

Notice that $\frac{1}{3}+\frac{1}{4}$ is greater than $\frac{1}{2}$. Why? Because $\frac{1}{3}$ is greater than $\frac{1}{4}$, so $\frac{1}{3}+\frac{1}{4}>\frac{1}{4}+\frac{1}{4}$, or $\frac{1}{2}$. But notice that $\frac{1}{5}+\frac{1}{6}+\frac{1}{7}+\frac{1}{8}$ is also greater than $\frac{1}{2}$ because $\frac{1}{5}$, $\frac{1}{6}$, and $\frac{1}{7}$ are all greater than $\frac{1}{8}$, so $\frac{1}{5}+\frac{1}{6}+\frac{1}{7}+\frac{1}{8}>\frac{1}{8}+\frac{1}{8}+\frac{1}{8}+\frac{1}{8}$, or $\frac{1}{2}$. The next group is also greater than $\frac{1}{2}$, as is the group after that. In fact, we can continue grouping terms in this way to produce an infinite number of groups, each greater than $\frac{1}{2}$. This means that our sum is *greater* than an infinite number of $\frac{1}{2}$s, and an infinite number of $\frac{1}{2}$s is most assuredly infinite. Therefore, the Harmonic Series is greater than infinity, forcing it also to be infinite.

Wait—can't we just do that same thing with $\frac{1}{2}+\frac{1}{4}+\frac{1}{8}+\frac{1}{16}+\frac{1}{32}+\cdots$? Can't we just group enough terms to get to $\frac{1}{2}$, and then group some more terms to get another $\frac{1}{2}$, and so on? Wouldn't that show that $\frac{1}{2}+\frac{1}{4}+\frac{1}{8}+\frac{1}{16}+\frac{1}{32}+\cdots$ is also infinite, even though we already showed that it was equal to 1? Fear not! The math is safe! The above technique of grouping numbers to be greater than $\frac{1}{2}$ does not work with $\frac{1}{2}+\frac{1}{4}+\frac{1}{8}+\frac{1}{16}+\frac{1}{32}+\cdots$. The first term is $\frac{1}{2}$, but to get to the next $\frac{1}{2}$ you would need *all of the rest* of the terms—*all infinitely many of them*. Not only is it impossible to get an infinite number of $\frac{1}{2}$s with $\frac{1}{2}+\frac{1}{4}+\frac{1}{8}+\frac{1}{16}+\frac{1}{32}+\cdots$, but it also turns out you can only get *two* of them, which is precisely why the sum is exactly equal to two halves, or 1.[46]

45 It takes 12,000 terms for the Harmonic Series to get above 10. It takes over 15 million trillion trillion trillion (or 15 with 42 zeros) terms for it to finally exceed 100.

46 The number $\frac{1}{2}$ is not in itself important to the argument. We might ask whether we can use a different number, $\frac{1}{3}$ for example, and see whether we can group terms together in just the right way to get an infinite number of $\frac{1}{3}$s. It turns out, though, that no matter the choice of number— $\frac{1}{2}$, $\frac{1}{3}$, or anything else—it will never work. This is precisely why the sum is finite.

Thus, sometimes the sum of an infinite number of infinitely small terms is infinite, and other times it is finite. It is as if the terms not only need to become smaller and smaller but also must become smaller *quickly*. If they take their good old time becoming tiny, then the sum can still blow up to infinity.

We will now use this idea of infinite sums and repeating decimals to prove one of the more counterintuitive facts in mathematics.

> *Repeating Nines Theorem*
> $0.\overline{9}=1$

This is a surprising result that has caused many an argument over the years in middle and high school mathematics classrooms. Even for some who have seen it, there is often still doubt about its veracity. How can 0.9999999999… actually equal 1? For this theorem, we offer not one but *three* arguments.

First, we use what is by now a familiar technique for dealing with repeating decimals. Consider the following subtraction problem.

$$\begin{array}{r} 9.9999999\cdots \\ -\ 0.9999999\cdots \\ \hline 9.0000000\ldots \end{array}$$

If ■ is our original number (0.9999999…), then this subtraction problem is

$$10\times■-■=9,$$

or

$$9\times■=9.$$

If we divide both sides by 9, we see that ■=1. Therefore, we have the astonishing fact that 0.9999999… = 1.[47]

The second argument starts with the fact that $0.3333333\cdots=\frac{1}{3}$. What happens if we multiply both sides of this equation by 3?

47 For those savvy with binary representations, this argument that $0.\overline{9}=1$ is exactly the same argument I gave for why $\frac{1}{2}+\frac{1}{4}+\frac{1}{8}+\frac{1}{16}+\frac{1}{32}+\cdots=1$. In binary, this sum is $0.\overline{1}$.

$$\begin{array}{rr} 0.3333333\cdots & = \frac{1}{3} \\ \times\ 3 & \times\ 3 \\ \hline 0.9999999\cdots & = 1 \end{array}$$

Finally, if it is still tempting to say that 0.9999999... is "very close to 1, but just shy of 1," the last argument is perhaps the best in refuting that objection. It relies on an important fact about real numbers.

> *The Density of the Real Numbers*
> Between two different real numbers, there is always another real number.

Why is this true? The easiest way to see that any two real numbers must have another number in between them is to use the *average* of the two numbers. After all, the average of two numbers always falls in between the two numbers themselves. If that is the case, we can always find a number in between *a* and *b* simply by averaging *a* and *b*, and then we can find more numbers in between those numbers, and so on. Truth be told, then, between any two real numbers there is not just one other number but an infinite number of other numbers.

Why is the density of the real numbers important? Why does it matter that any two non-equal numbers will always have a number in between them? Well, if 0.9999999... is less than 1, then there must be a number in between 0.9999999... and 1. What could that number possibly be? No matter how hard we try, it remains impossible to find one. To see this, let's look at how to increase a number by using one that is slightly less mysterious, say 0.12345. If we want to create a larger number while still staying less than 1, we have only two choices. The first is put another digit on the end: $0.12345\underline{6}$. The second is increase one of the existing digits: $0.\underline{7}2345$ or $0.12\underline{7}45$. Of course, if my decimal is infinitely repeating, such as $0.3333333\cdots$, then I cannot put a digit onto the end. I can, however, increase one of the existing digits: $0.33\underline{7}3333\cdots$ or $0.33333\underline{7}3\cdots$. The problem with 0.9999999... is that we can do neither of these things. We cannot put a digit onto the end because is it infinitely repeating, and we cannot increase any of the existing digits because they are all 9s. Therefore, there cannot be a number in between 0.9999999... and 1, and so we have no choice but to conclude that 0.9999999… must actually be 1, no more and no less.

CHAPTER 7
HOW MANY PRIME NUMBERS ARE THERE?

PRIME NUMBERS form perhaps the most interesting category of numbers in all of mathematics. The mere fact that they are interesting is in itself interesting, or at least surprising. To understand what I mean, let's first think about what a prime number actually is. Rather than jumping right into a formal definition, I offer a much more intuitive description.

> DEFINITION: The prime numbers
> are the basic building blocks of multiplication.

Numbers can be built using other numbers. At the earliest ages, we teach students to break numbers apart and put them back together. This sort of playful math is immensely valuable for students. True, we want young children to memorize math facts—failing to do so will cause unnecessary roadblocks for more advanced topics. *Number sense*, however, is far more important than memorized facts, though the two are related. Number sense is about establishing a certain fluency with numbers, knowing their parts and knowing how they combine with other numbers. Think of it this way. Memorizing math facts is about the ability to see the expression "9 + 8" and think immediately, "17." Number sense is about seeing the number "17" and thinking immediately, "9 + 8," and then thinking, "20 − 3," and then thinking, "This number is a prime number, and one more than a perfect square, and the average of two perfect numbers," and then countless other relationships and facts about the number 17. There is an analog concerning computation. Computation is about seeing "19 × 5" and being able to perform the standard multiplication algorithm. Number sense is about seeing "19 × 5" and thinking, "20 × 5 is 100, so 19 × 5 must be 5 less, or 95." Number sense begins by understanding that numbers are building blocks for other numbers, and this sense can only be developed by constantly *playing with numbers*.

When we have an operation, we have the possibility of basic building blocks. For example, let's consider the most fundamental operation: addition. What are the basic building blocks of addition? Before answering that question, I should clarify that our universe of numbers consists only of 1, 2, 3, 4, 5, ...—in other words, the whole numbers. Yes, it is true that

$1 = 0.2 + 0.8$, which does mean that the number 1 is built from 0.2 and 0.8, but this is out of scope for our current discussion. Using a construction analogy, if we are asked to identify the basic building blocks of a wall, we would likely answer "the bricks," even though the bricks can be broken down into their molecules and their molecules into atoms.

The same thing is true for our discussion about whole numbers. It is interesting in itself that numbers can continue to be broken down into smaller and smaller decimals, particularly because this means that in the universe of decimals *there are no basic building blocks*. However interesting that is, though, it is not the interesting question we are currently investigating. For similar reasons, we are also not interested at this point in using negative numbers in our quest to locate building blocks. If we break the number 8 into $10 + (-2)$, we could also end up with an infinite process of breaking apart, and therefore a lack of basic units. Finally, we also need to specifically exclude the number 0 as a candidate for a basic building block. This exclusion is for two reasons. First, it also would lead to an infinite breaking apart, as 6 could be $6 + 0$, or $6 + 0 + 0$, or $6 + 0 + 0 + 0$, etc. Second, 0 is a very special number for addition. In some weird way, it exists and can be added, but it does nothing. (Might we say it "adds nothing" to the result when added?) Breaking apart numbers using 0 is like being asked to cut a deck of cards and simply choosing not to cut. This choice is interesting, to be sure, and it shows a lot about the nature of the number 0 as well as the card player, but it is a choice *not* to cut rather than a particular kind of cut. Similarly, splitting numbers up with addition and the number 0 is not really splitting numbers up; it is choosing not to split.

We are now ready to consider our question: "What are the basic building blocks of addition for the universe of whole numbers?" One way to explore this is to start with any number and disassemble it. We choose, for no particular reason, the number 6. We can split 6 into 2 and 4, because $6 = 2 + 4$. This is not the only way to split it, but it is one way. It will be helpful for us to switch metaphors from building blocks to diagram we might call an "addition tree."

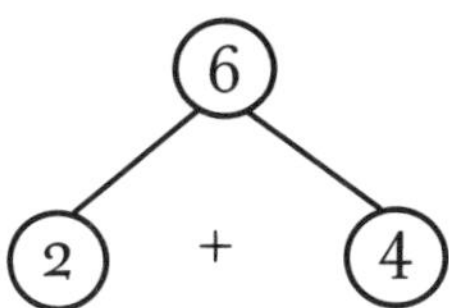

The numbers 2 and 4 can be further broken down by addition. Rather than showing each step one at a time, the tree below gives the complete breakdown of 6.

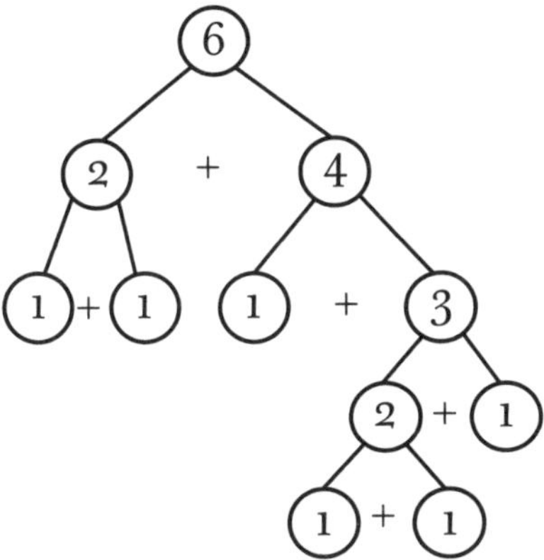

Why is this tree complete? Once we get to 1, there is no further breaking apart that can happen. In our current universe (no fractions, no negative numbers, and a prohibition against using the number 0), there is nowhere to go from 1.[48]

After all of that discussion and all the work to split apart the number 6, we have the stark reality that all the leaves of the tree are 1. In other words, no matter how we go about breaking up the number 6 into all of its additive parts, we will eventually get $6=1+1+1+1+1+1$. Moreover, the number 6 was not special, short of the fact that all numbers are special. *Any* positive whole number can be broken down into a sum of 1s. This is either fascinating or boring, depending on how we think about it: *the only basic building block of addition is the number 1.* In other words, all whole numbers can be expressed as a sum of 1s. The only number that cannot be broken down into parts, according to the rules of the addition tree, is the

48 The number 0 is known as the *additive identity* because adding it to any other number will result in the same number. Similarly, the number 1 is known as the *multiplicative identity* because multiplying it by any other number will result in the same number. It is an astonishing fact that there are no integers between the additive identity and the multiplicative identity. Can you prove this?

number 1. If we define "additive prime numbers" as "the basic building blocks of whole number addition," we would have only one "additive prime number," i.e., the number 1. For what it is worth, I will list this as a theorem, but one for which I do not have a creative name.

Copies of the Number 1 Theorem
Any whole number can be written as the sum of 1s.

This process of breaking numbers apart using addition gives us some ideas and tools for looking at the actual prime numbers. Prime numbers, as defined at the start of this chapter, are "the basic building blocks of *multiplication*." Using our other analogy, they are the leaves of multiplication trees, more commonly called "factor trees." As was the case with addition, our universe is the whole numbers. But there is one further stipulation that we need to make for multiplication. The number 1 cannot be a basic building block. This is for the same reason that 0 was disqualified for addition. In the case of addition, 0 leads to an infinite process of breaking up numbers: $6=6+0=6+0+0=6+0+0+0$, and so on. The number 0 does not really contribute anything to addition. The same thing is true for the number 1 when it comes to multiplication. If 1 were a basic building block, then we would have a similar infinite splitting: $6=6\times1=6\times1\times1=6\times1\times1\times1$, etc. In fact, the number 1 plays an identical role in multiplication to that of 0 in addition in that it contributes nothing to the operation. Mathematically, we call both of these "identities" with respect to their operation. The number 0 is the "additive identity" because any number plus 0 is itself. Similarly, 1 is the "multiplicative identity" because any number multiplied by 1 is itself. In both cases, they are excluded from being a basic building block, or a leaf of a tree.

We are now prepared to define prime numbers with a bit more precision.

DEFINITION: A *prime number* is a whole number greater than 1 that cannot be properly broken up into the product of two other whole numbers. (By "properly broken up" we mean using parts that are not equal to 1.)

Upon reading this definition carefully, we note that the number 1 is not a prime number. This is not a mistake, and to assert otherwise is one of the most common misconceptions in primary school math classrooms. Sometimes prime numbers are defined as "numbers that are only divisible by themselves and 1." The problem with this definition is that it makes 1 seem like a prime number, *and it is not!*[49] The whole spirit of the primes is to serve as basic building blocks for multiplication. We should be able to use them to construct other numbers. For example, we can construct 6 by multiplying 2 and 3 together. The number 1 is not a building block because it does not generate any new numbers using multiplication. We will see later in the Fundamental Theorem of Arithmetic a further reason why 1 should not be a prime number.

To break a number down into its fundamental blocks, we use the same tree structure that we used for addition, but this time we use multiplication. Consider the number 240, which can be split apart into 4×60. This is not the only way to split it up, of course, but it is a perfectly good way.

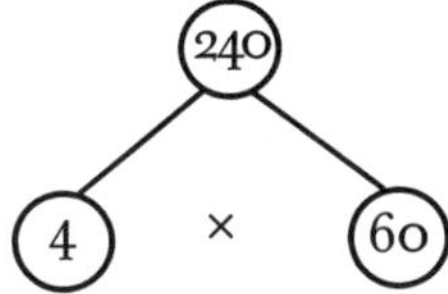

The numbers 4 and 60 can be further broken down by multiplication.

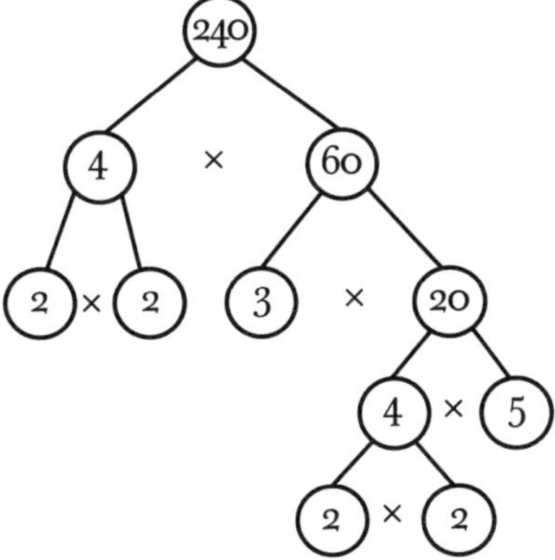

It is worth pausing to convince yourself that the numbers 2, 3, and 5

49 This does not mean that the number 1 is boring. On the contrary, it is quite fascinating, because 1 is also not a composite number. It is not *composed* of other numbers by multiplication, making it the only whole number that is neither prime nor composite.

cannot be broken down with multiplication any further. Once we construct the entire factor tree, we can write our original number as a product of these prime numbers: $240 = 2 \times 2 \times 3 \times 2 \times 2 \times 5$. There is a theorem of sorts that comes out of the construction of these factor trees that is a bit more interesting than its additive counterpart (the *Copies of the Number 1 Theorem*).

> *The Not-Quite-Fundamental Theorem of Arithmetic*
> Any whole number greater than 1 is either prime
> or it is the product of primes.

The theorem is true because of the process we used to grow factor trees. It actually says no more than that a factor tree is either a leaf itself or it can be branched off into leaves. With numbers, we keep breaking them up until we cannot break them up anymore. When that happens—when we are at a leaf—we have a prime number. We can then use these leaves to write the original number as a product of primes. The only way this would be impossible is if the starting number were prime to begin with so that its tree were simply a single leaf, in which case we wouldn't really have a product. In terms of factor trees, the Not-Quite-Fundamental Theorem means that any whole number greater than 1 is either prime or it has a factor tree that ends in primes.

We have skipped something that is rather amazing, though, and it is the reason I stated at the beginning of this chapter that the fact that prime numbers are interesting is itself interesting. Remember, when we tried to define the concept of "additive primes," we found that there was only one basic building block: the number 1. So far, we have only explored one number's factor tree, i.e., 240, and yet in this one example, we already have a delightful *three* different basic building blocks (2, 3, and 5)! What's more, it now raises the question, "How many of these basic building blocks are there for multiplication?" In other words, how many prime numbers exist? It is reasonable, perhaps, to assume that there are a small number of them, from which every other number is produced using multiplication. What if 2, 3, and 5 are the only three primes? Can we build up all numbers through multiplication using just these three numbers? We could just start exploring, of course. We could pick random numbers and break them down—and we likely *would* discover new primes—but it may be good to think about being more methodical.

What I propose we do is to look at the number $2\times3\times5+1$. Does $2\times3\times5+1$ have the number 5 in its factor tree? If you think carefully about this, the only numbers with 5 in their trees are multiples of 5, for example 5, 10, 15, 20, 25, 30, etc. But $2\times3\times5+1$ is *not* a multiple of 5. In fact, it is exactly 1 more than a multiple of 5, and a number that is 1 more than a multiple of 5 can never itself be a multiple of 5 because multiples of 5 are exactly five numbers apart.

So much for 5, but does $2\times3\times5+1$ have 3 in its factor tree? Again, the answer has to be no. This is because $2\times3\times5$ is not only a multiple of 5 but also a multiple of 3. Therefore, $2\times3\times5+1$ is 1 more than a multiple of 3, and a number that is 1 more than a multiple of 3 can never be a multiple of 3 because multiples of 3 are exactly three numbers apart. Finally, by this very same reasoning, $2\times3\times5+1$ cannot be a multiple of 2 because it is 1 more than a multiple of 2.

We now know that the factor tree for $2\times3\times5+1$ does not have 2, 3, or 5 in its leaves. Why is this significant? We know from the Not-Quite-Fundamental Theorem of Arithmetic that *any number*, including $2\times3\times5+1$, must either be *prime* or *the product of primes*. The number $2\times3\times5+1$ is not one of our known primes (2, 3, or 5), and it does not have one of these known primes in its factor tree. Therefore, $2\times3\times5+1$ is either a new prime number itself or there must be an undiscovered new prime number in its tree. In point of fact, the actual value is $2\times3\times5+1=31$. The number 31 *is* a prime number, so we have successfully discovered a fourth prime number.

Let's see whether we can generate another prime using our new list of 2, 3, 5, and 31. We will proceed as we did before. Consider the number $2\times3\times5\times31+1$. This number will not have 2, 3, 5, or 31 in its factor tree because it is one more than a multiple of each of these numbers. Because of the Not-Quite-Fundamental Theorem of Arithmetic, however, $2\times3\times5\times31+1$ must either be prime or be the product of primes. Calculated, the number is 931. This case is different than the first one because 931 is *not prime*. Rather than being a leaf without a tree, it has a proper factor tree with prime leaves at the end. But following our argument from before, these primes cannot be 2, 3, 5, or 31, so the leaves must be *new* primes. To find them, let's construct the full factor tree.

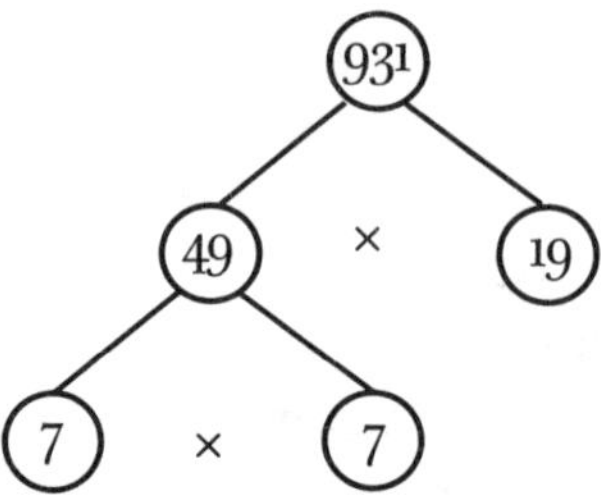

The fact that $2\times3\times5\times31+1$ is not itself a prime number doesn't matter. What does matter, though, is that we have *still* generated a new prime number—it's just that the new prime is not 931. In fact, we have generated *two* new prime numbers: 7 and 19!

Where do we go from here? We once again take all of our known primes (2, 3, 5, 7, 19, and 31), multiply them together, and add 1. The result either is a new prime number, or we will find new prime numbers in the leaves of its factor tree. The process can continue over and over again, generating prime after prime, *ad infinitum*,[50] which leads us to a beautiful and surprising theorem.

> *The Number of Primes Theorem*
> There are an infinite number of prime numbers.

Remember, when we looked for the basic building blocks of addition, we discovered that there was only one, namely the number 1. When looking for the basic building blocks of multiplication, we find out not only that more than one exists but that there are an infinite number of them! To appreciate this fact, note that even the basic building blocks of the physical universe are finite. That is, there are likely only a finite number of types of subatomic particles. Once again, we find mathematics transcending the material universe in its ability to contain and deal with the infinite.

The above proof is a version of the one offered by Euclid in the *Elements*. It may seem curious that the *Elements* is a book about geometry, but Euclid proves this theorem (and many others) about *numbers*. As I mentioned earlier, this is because for Euclid numbers really are shapes

50 It should be noted that it is not immediately clear whether this process will generate *all* the primes. For those who know something about primes, it will be clear that some of them have been skipped, at least at these early stages.

(magnitudes). His statement about prime numbers is really a statement about certain lengths of lines and how they combine in ratios.

What I have presented is something of a "constructive" version of Euclid's proof, beginning with three very particular prime numbers and describing a process that can be performed repeatedly to generate new primes. I would be remiss if I did not at least outline the more formal version of this remarkable proof.

We are trying to prove that there are an infinite number of prime numbers. We start by assuming the opposite—that there are a finite number of prime numbers—and we write them out: $p_1, p_2, p_3, \ldots, p_n$. Consider the number $M = p_1 \times p_2 \times p_3 \times \ldots \times p_n + 1$. This number cannot be prime because we said that we had all of the primes ($p_1, p_2, p_3, \ldots, p_n$), and M, which is 1 more than their product, is clearly greater than all of these primes. Because our new number, M, is not prime,[51] then it must be the product of primes. This product, however, cannot contain p_1 because M is one more than a multiple of p_1. In other words, there is a remainder of 1 when M is divided by p_1, so M is not a multiple of p_1.

Now, there was nothing special about choosing p_1. M is also one more than a multiple of p_2, and therefore, M's prime product cannot contain p_2 either. In fact, for the same reason M's prime product cannot contain *any* of the prime numbers $p_1, p_2, p_3, \ldots, p_n$. This is a serious problem. Why? Because M is not a prime number and yet cannot be broken up into any of our known prime numbers. This contradicts the Not-Quite-Fundamental Theorem of Arithmetic. Whenever we have a contradiction, we look back to see whether we made any unproven assumptions. The only assumption we made was that we had written down all the primes. In other words, we assumed that there was a finite number of primes to begin with. Because this led to a contradiction, that assumption must be false, and therefore there are, in fact, an infinite number of prime numbers.

51 It is a common mistake in presenting this proof to claim that M must be prime. This is not true, as we saw in our specific construction ($2 \times 3 \times 5 \times 31 + 1$ was *not* prime). Thankfully, it need not be true for the proof to work. The point is not that we have generated a new prime in M but rather that the factor tree for M must have a prime that was missed in our supposedly "complete" list of primes.

CHAPTER 8
THE FUNDAMENTAL THEOREM OF ARITHMETIC

WE NOTED earlier that the proof of the Pythagorean Theorem was the first difficult step on our journey. It was a step up from the Triangle Angle Sum Theorem, polygonal tilings, and the Platonic solids. The proof in this chapter is another step up and might be the most difficult among those offered in this part of the book. This difficulty starts with the number of parts needed to build up a proper proof, but even some of the steps within those parts will require us to pause. Nevertheless, at this point we have practiced our proof skills and honed our mathematical intuitions, and we are up for the challenge.

In the previous chapter we used factor trees to visualize prime numbers as the basic building blocks of multiplication and to prove a theorem we called the Not-Quite-Fundamental Theorem of Arithmetic.

> *The Not-Quite-Fundamental Theorem of Arithmetic*
> Any whole number greater than 1 is either prime
> or it is the product of primes.

The full Fundamental Theorem of Arithmetic is much stronger. It states that this product of primes is *unique*.

What does that mean? Consider the number 240, which we broke down into its prime parts.

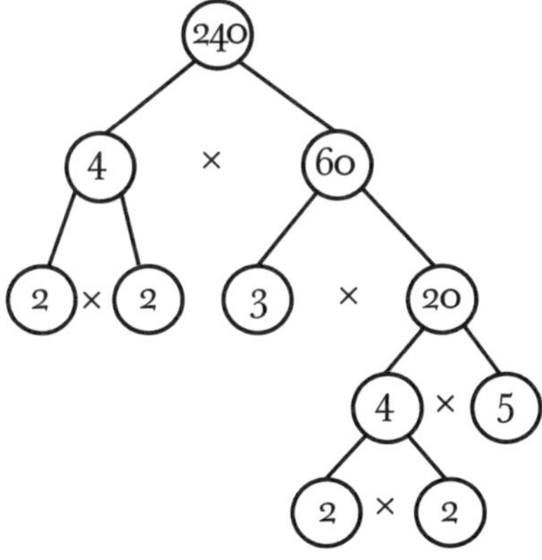

There is no reason that we had to start with 4×60. There are several ways we could have broken apart 240. For example, what would happen if we started our factor tree using 10 and 24?

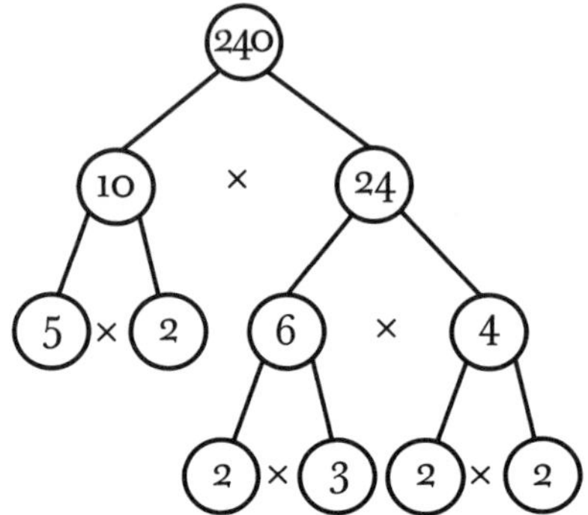

The tree is different from the original one, but notice that the numbers on the leaves are the same. In both we end up with four 2s, one 3, and one 5, or $2^4 \times 3 \times 5$. The Fundamental Theorem of Arithmetic says that this will always happen. No matter how we split 240 in the first step, and no matter how we split it during the subsequent steps, we will always end up with the same prime numbers. Only the *order* of the primes in the tree might differ. This is another reason why 1 is not considered a prime number. If 1 were prime, we would lose unique factorization. For example, the number 12 could be factored as $2 \times 2 \times 3$ but also as $2 \times 2 \times 3 \times 1$ and $2 \times 2 \times 3 \times 1 \times 1$, etc.

> *The Fundamental Theorem of Arithmetic*
> Any whole number greater than 1 is either prime or it is the product of primes, and its prime factorization is unique.

The first part of the theorem—that we can write a number as the product of primes—is what we called the Not-Quite-Fundamental Theorem of Arithmetic and was fairly intuitive. After all, at some point the factor tree has to stop. The fact that we will always get the *same* primes every time it stops is not at all intuitive, even though we likely saw example after example during our time in late elementary school. To help us navigate the proof, we are first going to present a lemma that is named after Euclid and appears as Proposition 30 in Book VII of the *Elements*.

Euclid's Lemma
If a prime number is a factor of $a \times b$,
then it must be a factor of either a or b.

Before proving this lemma, it is worth giving an example. The number 7 is a prime number, and it is a factor of 126. Moreover, $126 = 14 \times 9$. Euclid claims that 7 must be a factor of either 14 or 9. This is true, of course, because 7 is a factor of 14. For some numbers, it could be that the prime is a factor of both parts. For example, 5 is a factor of 375, and $375 = 25 \times 15$. It just so happens that 5 is a factor of *both* 25 and 15, which is fine. The lemma just says that the prime number must be a factor of at least one of the two.

In order to prove Euclid's Lemma, we start with a prime p that is a factor of a product $a \times b$. If p is a factor of a, then the lemma is satisfied, so let's consider the case in which p is *not* a factor of a. Because p is a factor of $a \times b$, then $a \times b$ must have a "matching" factor for p. (Factors always come in pairs. For example, the numbers 9 and 4 pair with each other to make 36. Sometimes a factor pairs with itself, as in the case of $6 \times 6 = 36$). Let's call the matching factor m, meaning that the quantity $a \times b$ can be factored as $p \times m$.

$$a \times b = p \times m$$

You may remember "cross multiplying" from various mathematics courses. The idea is that two equal fractions can lead to equal products.

$$\frac{3}{6} \quad \frac{4}{8}$$

$$6 \times 4 = 3 \times 8$$

The expression $a \times b = p \times m$ is already an equality of products, so we can "un-cross multiply" it, if you will.

$$\frac{b}{m} = \frac{p}{a}$$

The fraction $\frac{p}{a}$ is interesting. The number p is a prime number, and so it has only two factors: 1 and p. The number a can have any number of factors, but p is not one of them. (Remember, this case is the one in which we assumed that p is *not* a factor of a.) This means that the only *common* factor of p and a is 1, which also means that the fraction $\frac{p}{a}$ is in lowest terms. (If it were not in lowest terms, p and a would have a common factor other than 1 that could be divided out.)

Let's make a short digression and remember how we go about putting a fraction in lowest terms. We take the numerator and denominator and divide by the greatest common factor. For example, $\frac{6}{15}$ can be put into lowest terms by dividing both the numerator and the denominator by 3.

$$\frac{6 \div 3}{15 \div 3} = \frac{2}{5}$$

Because all factors come in pairs, when we divide the numerator by a factor, we obtain the matching factor in the new numerator. Likewise, when we divide the denominator by a factor, we obtain its matching factor in the new denominator. In our example, 3 was the greatest common factor of 6 and 15. When we divided 6 by 3, we got 2, which is another factor of 6. Similarly, when we divided 15 by 3, we obtained 5, which is another factor of 15. Therefore, when a fraction is put into lowest terms, the reduced numerator is a factor of the original numerator, and the reduced denominator is a factor of the original denominator.[52]

Reduced Fractions Lemma
When a fraction is reduced to lowest terms, the reduced numerator is a factor of the original numerator, and the reduced denominator is a factor of the original denominator.

When we look back at the "un-cross-multiplied" equation from above, $\frac{b}{m} = \frac{p}{a}$, we have written the fraction $\frac{b}{m}$ in lowest terms as $\frac{p}{a}$. By the Reduced Fractions Lemma, it must be that p is a factor of b. (It must also be that a is a factor of m, but that is not as relevant for us.) Therefore, we have proven Euclid's Lemma. If p is a factor of $a \times b$, either p is a factor of a or, if not, then by the above argument p must be a factor of b.

It will not take us long to extend Euclid's Lemma to three factors. If p

52 This property is Proposition 20 in Book VII of Euclid's *Elements*.

is a factor of $a \times b \times c$, then it must be a factor of at least one of a, b, or c. Why? This is because $a \times b \times c$ can be written as a product of two numbers: $a \times (b \times c)$. Euclid's Lemma says that p must therefore be a factor of one or the other, either a or $(b \times c)$. If p is not a factor of a, then it must be a factor of $(b \times c)$, but then Euclid's Lemma applied again says that p must either be a factor of b or a factor of c. Of course, we can then move from three to four factors, and so on.

> *Extended Euclid's Lemma*
> If a prime number is a factor of a product,
> it must be a factor of one of the members of that product.
> In other words, if a prime number is a factor of
> $a_1 \times a_2 \times \ldots \times a_n$, then it must be a factor of at least one a_i.

We are now ready to complete the proof of the Fundamental Theorem of Arithmetic. The task is to show that the prime factorization of a number is unique. To show this, we will suppose that we actually have *two* factorizations of a number and try to show that these two factorizations are really the same, differing only perhaps by the order of the primes.

$$N = p_1 \times p_2 \times \ldots \times p_n$$
$$N = q_1 \times q_2 \times \ldots \times q_m$$

Take any prime factor from the first factorization, say p_1. Because p_1 is a factor of N, and because $N = q_1 \times q_2 \times \ldots \times q_m$, the Extended Euclid's Lemma says that p_1 must be a factor of at least one q_i. But every q_i is also prime, and the only way for one prime to be a factor of another prime is if the two primes are equal. Therefore, p_1 must be the very same prime as q_i, and so each factorization has at least one prime factor in common. We can divide by that prime number in both factorizations to eliminate it, and then repeat the process, finding another pair of identical primes, and then another, and then another, until all of the primes have been divided out. Therefore, each p_j has a matching q_i, and vice versa, meaning that the two lists are identical, with the only exception being the order in which they are written. The factorizations are the same, and we now have a proof of the Fundamental Theorem of Arithmetic.

We stated at the beginning that the Fundamental Theorem is not entirely obvious, even though we have known it since elementary school. One argument for its non-obviousness is the difficulty of the proof. Another argument is that unique factorization does not exist in other systems. For the mathematically curious, there is a system that consists entirely of numbers of the form $a+b\sqrt{-5}$,where a and b are integers. In this system, the number 6 can be factored into primes in two entirely different ways.

$$6=2\times 3=(1+\sqrt{-5})\times(1-\sqrt{-5})$$

It can be shown in this system that 2, 3, $1+\sqrt{-5}$ and $1-\sqrt{-5}$ are all prime numbers, though that would require some careful thinking about the meaning of "prime" in this new context.[53] This system of numbers has prime factorization but not *unique* factorization.[54] Knowing this makes it at least a little surprising that the factorization of whole numbers is unique, and perhaps we have a glimpse into why this theorem is so "fundamental" for normal arithmetic.

53 The more general definition of a prime number that can apply to systems other than the whole numbers is "a number that has no proper factorization." That is, if the prime number is written as a product, one of the two factors must be a unit. In the whole numbers, the only unit is 1. In the integers, the two units are 1 and -1. Therefore, the number -7 is considered to be a prime number in the integers. For the system consisting of numbers in the form $a+b\sqrt{-5}$, there are four units: 1, -1, $\sqrt{-1}$, and $-\sqrt{-1}$. Moreover, a unit is never considered a prime number.

54 It is an astounding result of mathematics that the only negative values of n for which numbers of the form $a+b\sqrt{n}$ have unique factorization are -1, -2, -3, -7, -11, -19, -43, -67, and -163.

CHAPTER 9
HOW MANY IRRATIONAL NUMBERS ARE THERE?

THERE IS a summer program for high school students at Ohio State called the Ross Mathematics Program, named after its late founder, Dr. Arnold Ross. The motto of the program is "Think deeply about simple things." Sometimes the very foundations of mathematics are seemingly simple but have incredible depth if we have the eyes to see them and the will to wonder about them. One of the earliest exposures to mathematics is when a parent asks a young child something like, "How many circles do you see?"

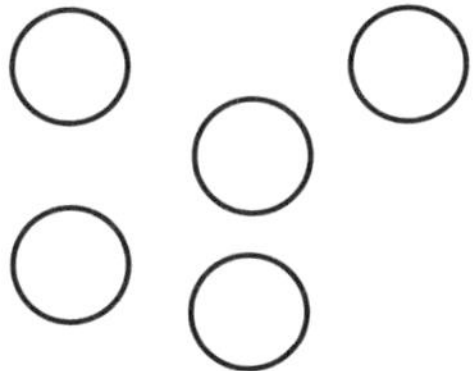

On the surface, this question seems simple. But have you ever stopped to think about what it *means* to say that there are five circles? As is often the case in life, we can get a hint of the deeper truth by paying close attention to young children. In this instance, we can watch the way they count objects. They point to each circle, being careful not to repeat or skip any. As they point, they recite the sequence of numbers "1, 2, 3, …" The last number spoken when running out of circles is the answer to "How many?" In other words, it is a matching game.

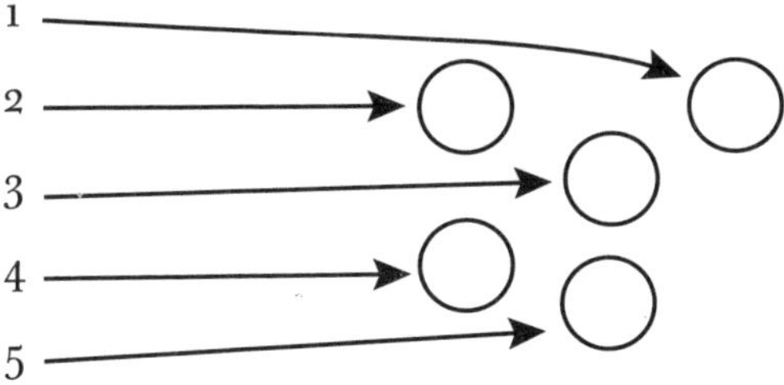

When mathematicians define the size of a collection, they do it using this same simple idea.

> DEFINITION: A collection has n things in it if there is an exact match between the objects in the collection and the numbers 1, 2, 3, … , n.

We can also use this idea to define when two sets have the same size.

> DEFINITION: Two finite collections have the same size if there is an exact match between the objects in the first collection and objects in the second collection.

For example, there are the same number of circles as squares because we can match them in such a way that every circle has exactly one square partner and every square has exactly one circular partner.

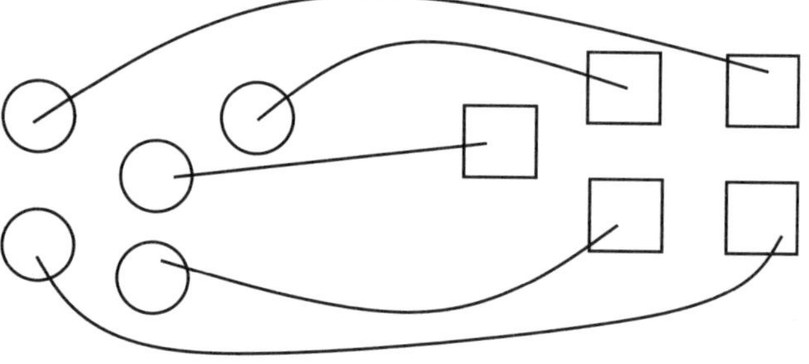

It is worth noting that these two definitions really are very close. In the first definition, the numbers 1, 2, 3, …, n can be thought of as objects in a set. The only difference is that the first definition allows us to label the size of a set with a number.

Why are we making a big deal about this? After seeing several proofs for the Pythagorean Theorem, after struggling through the algebra involved in showing that $\sqrt{2}$ is irrational, and after working our way through infinitely repeating decimals, why are we pausing on such an elementary topic as counting objects? Well, sometimes when we pause to think deeply about simple things, we can apply those same ideas to other things that seem more complicated. In this case, when we pause to think about *smaller* things, we can extend these ideas to *larger* things—in fact, *infinite* things.

Infinite sets are bizarre; there is no way around it. Strange yet wonderful things start happening when we play with infinity. The first exposure young people have to infinity might be in some sort of argument that involves trying to come up with larger and larger numbers. A dialogue might begin with two children, Jane and John, with Jane claiming, "I will be rich someday. I will have one million dollars!"

"Oh yeah?" says John. "I will be richer. I will have one billion dollars!"

"Well," returns Jane, "then I will have one trillion dollars."

Sensing victory, John says, "Then I will have *infinity* dollars."

Jane thinks she can one-up this (literally) and retorts, "Then I will have *infinity plus 1*."

Her opponent says, "That is still infinity. You haven't beaten me. Nothing beats infinity, so I win."

Is John correct that "infinity plus 1" is still infinity, and moreover that "nothing beats infinity"? The first question is easier, so let's start there. On the surface of it, John's argument seems reasonable. If you have an infinite amount, adding 1 to it means you still have an infinite amount. "Not so fast," says the mathematician. "Are the two amounts *the same*?" This question is at the heart of John's first claim that "infinity plus 1 is still infinity." For example, do the following two infinite collections have the same number of elements?

1, 2, 3, 4, 5, 6, 7, 8, 9, 10, 11, 12, 13, 14, 15, 16, 17, ...

2, 3, 4, 5, 6, 7, 8, 9, 10, 11, 12, 13, 14, 15, 16, 17, 18, ...

All of a sudden, this question is interesting. On the one hand, the first collection seems to have more elements—precisely *one more*, in fact. It has every element from the second collection, plus the element 1. On the other hand, isn't infinity simply infinity no matter how we slice it—or add to it? It turns out that the conversation between the two children about "infinity plus 1" deserves some scrutiny. We could, of course, define one infinite collection to be larger than a second infinite collection if it has all the elements of the second collection and at least one extra. The problem with this definition is that it will not allow us to compare collections that don't share elements, such as the positive integers (1, 2, 3, 4, ...) and the negative integers (−1, −2, −3, −4, ...). The two collections have no overlap. What would your intuition say about the positive integers and the negative integers? Does one set have more elements, or are the two sets the same size? A reasonable answer seems to be that they *are* the same size. After all, although the elements do not overlap, strictly speaking, they are *almost* identical. The first set has the number 1, and the second set has the number −1. Similarly, the first set has the number 2, and the second set has the number −2. We can continue from here. If you think about it, this approach is the same that we used to compare finite sets. We knew that the number of circles was equal to the number of squares because we were able to provide exact one-to-one matches with no shape left out. The same thing is going on here. We can match the positive integers to the negative integers with every number having exactly one partner.

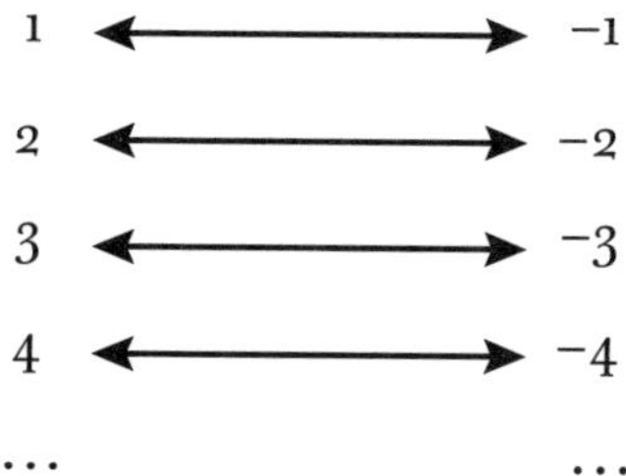

Of course, it is not good enough just to write down four pairs. We need to convince ourselves that *every* negative number, all infinitely many of them, really does have a positive partner, and vice versa. We need to write down a "rule" of pairing. In this case, it is not terribly difficult—every positive number matches exactly to its negative counterpart, and every negative number matches exactly to its positive counterpart. Using this example, we can formally adapt our "finite" version of "same size" to the infinite case.

> DEFINITION: Two infinite collections have the *same size* if there is an exact match between the objects in the first collection and objects in the second collection.

With this nice definition in place, we return to the question, "Do the following two collections have the same number of elements?" Are they the same size?

$$1, 2, 3, 4, 5, 6, 7, 8, 9, 10, 11, 12, 13, 14, 15, 16, 17, \ldots$$

$$2, 3, 4, 5, 6, 7, 8, 9, 10, 11, 12, 13, 14, 15, 16, 17, 18, \ldots$$

To answer this question, we should see whether we can match them up.

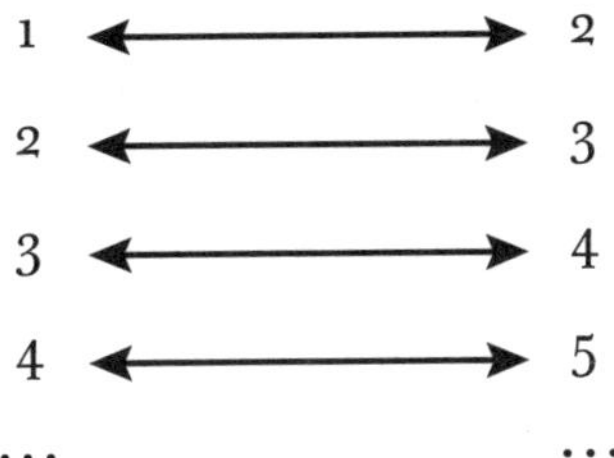

Here is where writing down a rule becomes important. It is one thing simply to list out some examples and draw some arrows between the numbers in each set. It is another thing altogether to convince ourselves that *every number on both sides of the arrows has exactly one match*, especially because we cannot actually write down all of the numbers. What would you say if asked, "How could you predict the matches between left and right?" If we look carefully, we can see that every right-hand number matches to "one fewer" than itself, and every left-hand number matches to "one more" than itself. We might describe the matching with a two-way machine.

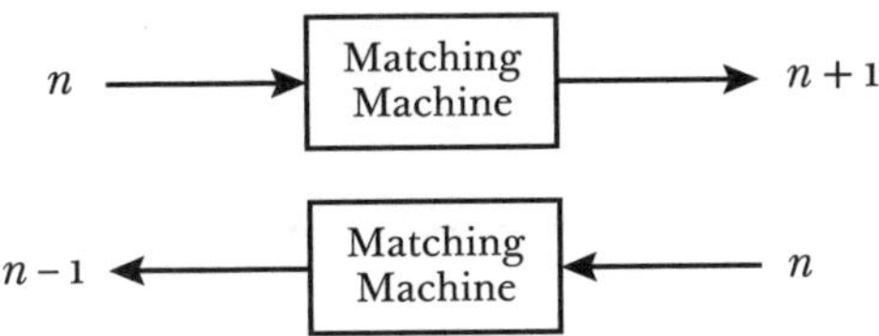

This matching machine is sufficient to convince us that every number in each set has one match and only one match, and therefore, the two sets have the same size. John was correct: "infinity plus 1" is still infinity. More accurately, an infinite set plus an extra element has the same size as the original set.

John made a much stronger assertion, however. He claimed that "nothing beats infinity." In the quest to see whether Jane can best John, what if we don't just add 1 to infinity, nor do we add 2, 10, or even 1,000,000,000 to infinity, but instead add *infinity* to infinity? Would "infinity plus infinity," what we might also call "infinity times 2," create something larger than just plain old infinity? To help us explore this, it would be good to think of a specific example. Which collection has a greater size, the natural numbers or the integers?

Natural Numbers: 0, 1, 2, 3, 4, 5, 6, 7, 8, 9, 10, ...

Integers: ..., −4, −3, −2, −1, 0, 1, 2, 3, 4, ...

It seems this time that the second collection is certainly larger than the first. In fact, it really does seem "twice as big." After all, the first set has two "copies" of the second set: the positive version and the negative version. On the other hand, we have already seen that some pretty weird things happen when we deal with infinity. To check their relative sizes, consider the following matchup:

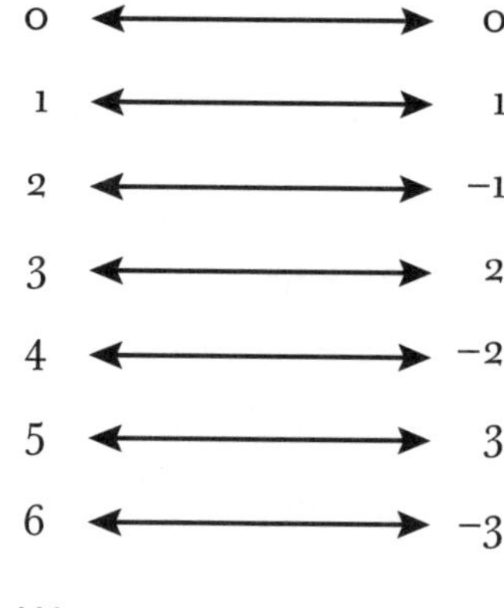

Just as before, it isn't enough simply to write down these seven pairs and claim that everything has a match. The matching must be predictable enough to convince us that *every* natural number that would ever appear on the left would have a matching integer on the right and that *every* integer on the right would have a matching natural number on the left. The key to seeing a natural number's match is to consider the even numbers and the odd numbers separately. The even numbers on the left will match to "half their value, but negative" on the right. So, for example, 2 matches with $-(2 \div 2)$, or -1, and 6 matches with $-(6 \div 2)$, or -3. What about the odd numbers? They match with positive integers, also in a predictable way. They match with "the value, plus 1, then divided by 2." For example, 3 matches with $(3+1) \div 2$, or 2, and 5 matches with $(5+1) \div 2$, or 3. This description might seem cumbersome, but it allows us to predict the match for any number on the left. For example, what is the match for 102? This number is even, so it will match with $-(102 \div 2)$, or -51. What about the number 255? This number is odd, so it will match with $(255+1) \div 2$, or 128. If you write out enough numbers, you can confirm these particular matches.

What about matching the other way? How can you predict the match for an integer on the right? Again, we can consider two cases separately—positive integers and negative integers—and then match 0 with 0. The two-way matching machine diagram would be a bit more complicated than our earlier examples, but it can still be described.[55]

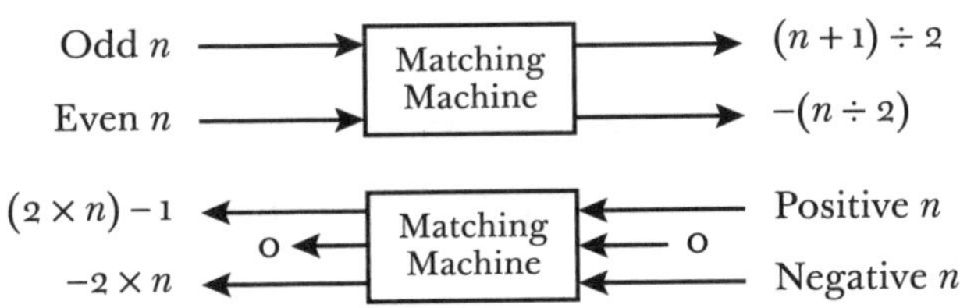

This matching machine is enough to convince us that every number on either side has one and only one match on the other side. Therefore, the two infinite collections really are the same size, which means that "two times infinity" (or "infinity plus infinity") really is "still infinity." More precisely, "Two times an infinity is equal to an infinity of the same size."

The situation is not looking good for Jane. It is starting to seem like all infinities might be the same size. Is there a response that Jane can offer? Can we find an infinity that is somehow larger than the ones with which we have been dealing? In other words, is there a set that is larger than the natural numbers? One candidate could be the rational numbers, or

55 It is worth noting that we really need the two directions to be inverses of one another. In other words, if sending 255 into the machine from the left gives us 128, then we need 128 to be sent back to 255 when coming from the right. In our case, if n is odd on the left, it is sent to $(n+1) \div 2$. The result will be a positive number, so when it is sent back from the right, it will go to $2 \times [(n+1) \div 2] - 1$, which simplifies back to n. This is important. I will let you work out the other cases.

the fractions. After all, *even just between the numbers 0 and 1, there are infinite number of rational numbers*: $\frac{1}{2}$, $\frac{1}{3}$, $\frac{2}{3}$, $\frac{9}{10}$, etc. There are another infinite number of rational numbers between 1 and 2, and again between 2 and 3, and so on. In fact, between any two natural numbers we can find an infinite number of rational numbers. It is like the rational numbers are an "infinity of infinities." Surely this is likely to be larger than just infinity. If so, Jane could claim, "Oh yeah, I will have *infinity times infinity* dollars," or alternatively, "I will have *infinity squared* dollars."

The German mathematician Georg Cantor developed the method we are using to compare infinite collections, and it is through his work that we have a remarkable and surprising result, albeit a disappointing one for Jane. *The set of rational numbers is the same size as the set of natural numbers.* To show this, we are going to focus only on the positive numbers from both collections. The task is to develop a strategy for matching the numbers 1, 2, 3, 4, 5, ... with *all* of the positive fractions. It amounts to being able to "list" the rational number in some organized way. This is how Cantor accomplished that task.

Numerator

Denominator

	1	2	3	4	5	6	...
1	$\frac{1}{1}$	$\frac{2}{1}$	$\frac{3}{1}$	$\frac{4}{1}$	$\frac{5}{1}$	$\frac{6}{1}$	...
2	$\frac{1}{2}$		$\frac{3}{2}$		$\frac{5}{2}$		...
3	$\frac{1}{3}$	$\frac{2}{3}$		$\frac{4}{3}$	$\frac{5}{3}$		...
4	$\frac{1}{4}$		$\frac{3}{4}$		$\frac{5}{4}$		...
5	$\frac{1}{5}$	$\frac{2}{5}$	$\frac{3}{5}$	$\frac{4}{5}$		$\frac{6}{5}$	...
6	$\frac{1}{6}$				$\frac{5}{6}$		...
...	...	...	...	...	...	...	...

Every rational number will appear in the chart because a rational number is a fraction formed by a numerator and a denominator, all of which are listed in the header row and header column. The gray spaces represent would-be duplicates. For example, one of the gray spaces would have been the fraction $\frac{2}{4}$, but this is the same as $\frac{1}{2}$, which is already represented. What is so ingenious about this? After all, it is merely a grid and not a matching machine. Cantor's brilliant move was to describe a strategy for traversing this grid so that every rational number is hit exactly once.

Numerator

Denominator	1	2	3	4	5	6	...
1	$\frac{1}{1}$	$\frac{2}{1}$	$\frac{3}{1}$	$\frac{4}{1}$	$\frac{5}{1}$	$\frac{6}{1}$	...
2	$\frac{1}{2}$		$\frac{3}{2}$		$\frac{5}{2}$		...
3	$\frac{1}{3}$	$\frac{2}{3}$		$\frac{4}{3}$	$\frac{5}{3}$		...
4	$\frac{1}{4}$		$\frac{3}{4}$		$\frac{5}{4}$		...
5	$\frac{1}{5}$	$\frac{2}{5}$	$\frac{3}{5}$	$\frac{4}{5}$		$\frac{6}{5}$	...
6	$\frac{1}{6}$				$\frac{5}{6}$		...
...	...	...	...	...	...	...	...

The matching scheme between the rational numbers and the whole numbers is described by the order in which the rational numbers appear in this diagonal tracing, skipping over the gray spaces.

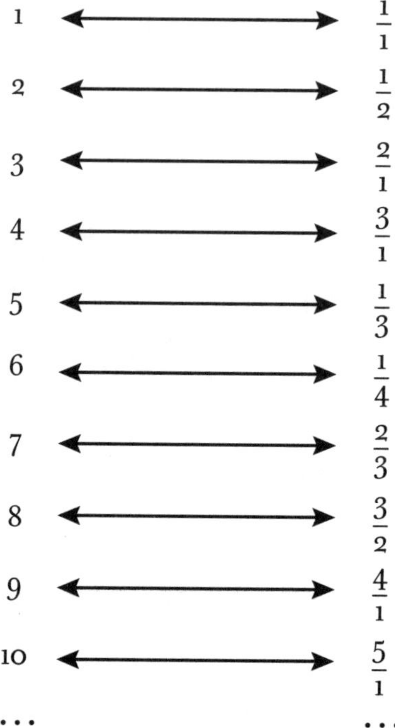

Given any left-hand number, the right-hand match is the rational number that is hit after that many steps in the diagonal pattern. Given any right-hand rational number, its left-hand match is how many steps it takes to hit the number in the diagonal pattern.[56]

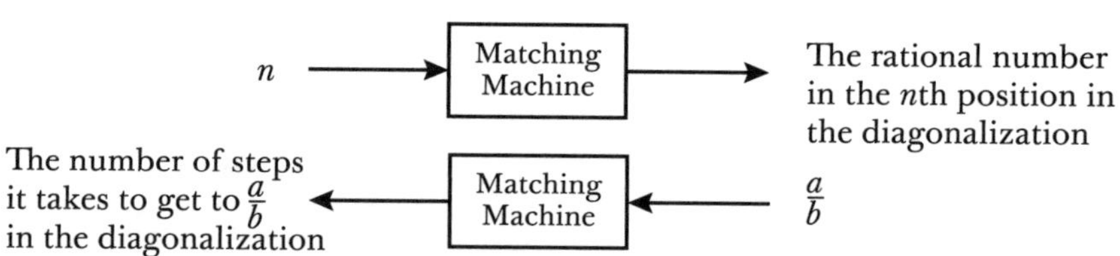

Remarkably, then, there are the same number of positive rational numbers as there are positive integers. We can then map the negative integers onto the negative rational numbers in exactly the same way. For example, whereas 6 and $\frac{1}{4}$ are matched in the above diagonalization, so too would

56 A formula actually can be worked out to map a natural number to a rational one and vice versa using this diagonalization, but it is quite difficult. The intuitive argument that it can be done is sufficient for our purposes.

be -6 and $-\frac{1}{4}$. Finally, we can match the integer 0 to the rational number 0, and then every rational number has exactly one integer match, and vice versa. Therefore, the number of rational numbers is the same as the number of integers.

> *Equal Infinities Theorem*
> The set of natural numbers, the set of integers, and the set of real numbers all have the same size.

Mathematicians say that these collections are *countable* because, even though they have an infinite size, we can still "count" them in the sense of naming the "third element" or the "three-hundredth element." This is not altogether different from how we count finite collections, with the notable exception that the counting never ends in the case of an infinite collection.[57]

The good news in all of this seems to be for John, who has yet to be defeated. Since he was the first to claim "infinity," we have not been able to beat him with "infinity plus one," nor with "two times infinity." We now see that we cannot even beat him with "infinity times infinity." Unfortunately, it also means that this whole theory of comparing sizes of infinity runs the risk of reaching a boring end. Have we wasted a lot of ink describing how to compare different "infinities" only to find out through several detailed examples that all infinities are the same size? Perhaps it is true that infinity really is *just infinity*. Perhaps we cannot get any larger, which was probably your intuition all along. It was certainly John's.

To rescue Jane, along comes Cantor, who accomplished one of the most outstanding feats in all of mathematics. He beat infinity. Or at least he beat the infinity that describes how many natural numbers there are. Cantor knew well that the infinity measuring the size of natural numbers does not become larger by moving to the integers, and it does not even become larger by moving to the rational numbers. But it *does* become larger by moving to the *real numbers*. The set of rational numbers is big, to be sure. There are a lot of them, an infinite number in fact, but there are not so

57 Note that a set's being countable is different from its being well ordered. For example, using the standard *less than* relationship, there is no "next" rational number after $\frac{2}{3}$ in the same sense that the "next" integer after 13 is 14. But the rational numbers can be *put* into a well-ordered list (even though that list is not governed by *less than*) so that $\frac{2}{3}$ does have a *next* number. Using our diagonalization, the *next* number is $\frac{3}{2}$.

many of them that we cannot "count" them, or rather "list" them. The real numbers, on the other hand, are not even listable; they are "*un*countable."

To prove this, Cantor gives a proof by contradiction. We first saw a proof by contradiction when showing the existence of at least *one* irrational number, namely $\sqrt{2}$, and then again in showing the existence of an infinite number of primes. We will use the same proof technique to show the existence of *a lot more* real numbers than there are integers or rational numbers.

We will start, actually, by showing that there are even a lot more *positive real numbers just between 0 and 1* than there are positive integers. Our goal is to show that these numbers can never be listed or counted by matching them up with 1, 2, 3, 4, ... We start by trying to do just that—assuming that we *can* match them in some way with 1, 2, 3, 4, ... It doesn't matter what matching we use, but suppose we have something like this.

1 ⟶ 0.349817349817340981730948710938470193874109 83...

2 ⟶ 0.980870984750987430598273409587098709587430 87...

3 ⟶ 0.113491230498170293875098899884354593450509 89...

4 ⟶ 0.234907834297084379805498753407984907852349 01...

5 ⟶ 0.431981759487509183470598170435981709587109 82...

6 ⟶ 0.333344804058549043905935450000000000000000 00...

7 ⟶ 0.121212121212121212121212121212121212121212 12...

Note that some of these numbers are terminating decimals, such as the sixth one, and some are repeating decimals, such as the seventh one. Others, however, are irrational numbers and have nonrepeating, nonterminating decimals. Somewhere in our list is the irrational number $\sqrt{2}$. The point is that we *claim* to have listed all of the real numbers between 0 and 1, both rational and irrational. The argument that this is impossible is so elegant that it assuredly deserves a place in *The Book*. We are going to construct a real number that cannot possibly be in this matched list even though we claim to have them all.

All real numbers between 0 and 1 have the same decimal form.

0._ _ _ _ _ _ _ _ ...

That is, all real numbers between 0 and 1 start with 0, then have a decimal point, and finally have an infinite number of digits. This is clear for irrational numbers and rational numbers that have repeating decimals, but even the terminating rational numbers take this form—they just have an infinite number of 0s after a while. The task is to construct a number that is guaranteed to be different from all the numbers in our list. If we can accomplish this task, we will have constructed an "unlisted" number, and this would contradict our claim that all numbers have been listed. There is a creative but simple way to find such an unlisted number.[58]

1. Make sure the first decimal place is different from the first number's first decimal place.
2. Make sure the second decimal place is different from the second number's second decimal place.
3. Make sure the third decimal place is different from the third number's third decimal place.
4. And so on.

In other words,

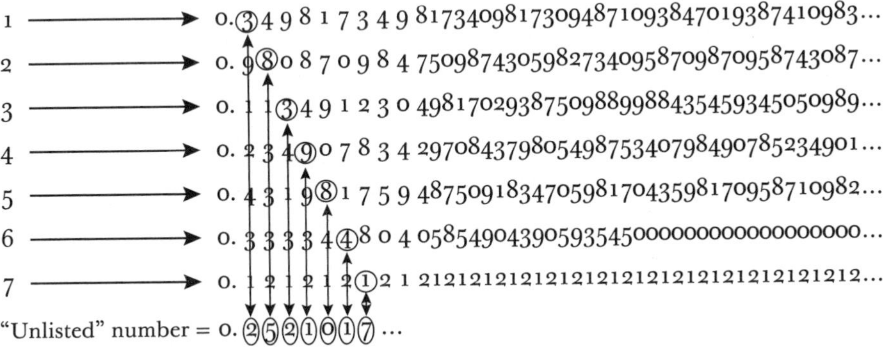

58 We have to deal with a small detail. From our previous work on infinite decimals, we know that some numbers can be written in two different ways. For example, the number 0.500000... can be written as 0.499999.... This does not invalidate our argument. We will just agree to write any terminating rational number with an infinite number of 0s rather than its other form involving an infinite number of 9s.

The particular replacement digit we choose for each place value doesn't matter as long as it is different from its counterpart.[59] How can we convince ourselves that our new number really is "unlisted," that it does not appear in our list, which claimed to have all real numbers (between 0 and 1)? Well, let's ask, "Is the new number the first number on our list?" The answer is no because the first number has a 3 in the tenths place, and the new number has a 2 in the tenths place. That was by design: we intentionally made the tenths place *different* from 3. We ask next, "Is the new number the *second* number in the list?" The answer again is no because the new number has a 5 in the hundredths place and the second number has an 8 in the hundredths place. This continues: the new number cannot be the third number, or the fourth number, and so on. We created a number that was intentionally different from every number in the list. This contradicts the fact that we claimed to have listed *all* the real numbers (between 0 and 1), but we then used that list to create a real number that was missed. True, we could add our new number, 0.2521017..., to the list, but then we could just run the process again to create another missed number. No matter how hard we try, we can always come up with a number that is not listed, even though we claim to have listed them all.

We used similar reasoning when showing that there are an infinite number of prime numbers. We assumed there were a finite number of primes and then used those primes to create a new number, thus demonstrating that we must have missed a prime number. The contradiction in that proof came from assuming the prime numbers were *finite*. The fundamental problem for the current proof is our assumption that the real numbers could be *listed* or *counted* or *matched up with the positive integers*. It is the listing of them that caused the problem.

One objection could be that we assumed a particular matching. After all, we only need to find one creative matching to show that a collection is the same size as the positive integers. Maybe there is some other listing we haven't thought of that would still allow the positive integers to be matched with all of the real numbers (between 0 and 1) without missing any. This is a very important point and a very good objection. But the

59 There is a second small detail that needs to be addressed. In making sure that our new decimal places differ from their counterpart, we will also never choose a 9. Therefore, we avoid any possibility of an infinite number of 9s in our newly created number, which could then appear in our original list under its "infinite number of 0s" form. This detail and the previous one are crucial for an exact proof, but I felt that they were too much to include in the main narrative.

argument we gave for creating a "missed" number did not at all depend on the listing we used. We illustrated a specific listing to help us understand the argument. *No matter what listing we try, no matter how creative we are, once we list them, we can always create an unlisted number.*[60] The problem was not the particular listing but the very act of listing itself.

At this point we have successfully completed the quest whose possibility we may have started to doubt: to find an infinity larger than the positive integers. Adding one element ("infinity plus one") failed. Doubling to get the set to the integers ("infinity times two") failed. Even moving to the rational numbers ("infinity of infinities" or "infinity times infinity"), which seemed at first to be clearly larger, failed. At last, however, we have conquered this "smaller" infinity.

> *The Size of (Some of) the Real Numbers Theorem*
> There are more real numbers between 0 and 1 than there are integers. The real numbers between 0 and 1 are uncountable, or "not listable."

How is Jane to respond to John to reclaim victory? Because real numbers between 0 and 1 are of the form 0._ _ _ _ ..., we can "calculate" the number of possibilities. There are ten choices for the first decimal place (0, 1, 2, 3, 4, 5, 6, 7, 8, or 9), ten for the second, ten for the third, and so on. Therefore, the "total" number of possibilities is $10 \times 10 \times 10 \times \ldots$, or 10^{∞}. Jane could say, "When I grow up, I am going to have $\$10^{\infty}$, which is *more* than your $\$\infty$!"[61]

Even though our quest, or at least Jane's quest, has ended, there are a few more points to address. We have shown that there are more real

60 It is worth considering why this argument does not also show that the rational numbers are uncountable, which would be a problem since we just showed they are countable. After all, rational numbers also have decimal expansions. Why can we not use the listing from the diagonalization, write down the decimal expansions, and then create a missed "unlisted" number? We can, actually. The problem is that the missed number is not guaranteed to be rational. In fact, quite the opposite is the case: it is guaranteed to be irrational. The key thing in the above proof is not simply that a "missed" number was created but that the number is also a *real number*, which is the set with which we are working.

61 There are two things to note here. First, in counting the real numbers this way, we again have the issue of counting both 0.500000... and 0.499999..., when in fact these are the same number. But this doesn't matter for Jane. Jane is simply trying to beat John's infinity. Because she overcounted, her infinity (10^{∞}) might be larger than the number of real numbers between 0 and 1, which we already showed, thanks to Cantor, was larger than John's infinity. Thus, she still emerges victorious. Truth be told, though, the overcounting doesn't contribute enough to create an infinity larger than the real numbers. Second, she doesn't even need to go to 10^{∞}. Even 2^{∞} would do the trick. That said, it is worth pointing out that there are, in fact, an infinite number of infinities. It is beyond the scope of this book, but if John and Jane were adventurous enough mathematicians, they could get into a battle of infinities that is, well, infinite. "I'll see your infinity and raise 2 to it!"

numbers between 0 and 1 than there are natural numbers, integers, and rational numbers. But because the entirety of the real numbers contains those between 0 and 1, this also means that the real numbers are a larger infinity than the smaller infinity that describes the rational numbers.[62]

> *The Size of (All of) the Real Numbers Theorem*
> There are more real numbers than there are natural numbers, integers, or rational numbers.

In other words, the rational numbers, while infinite, are the "smaller" kind of infinity, and they are included in the "larger" infinity that is the real numbers.

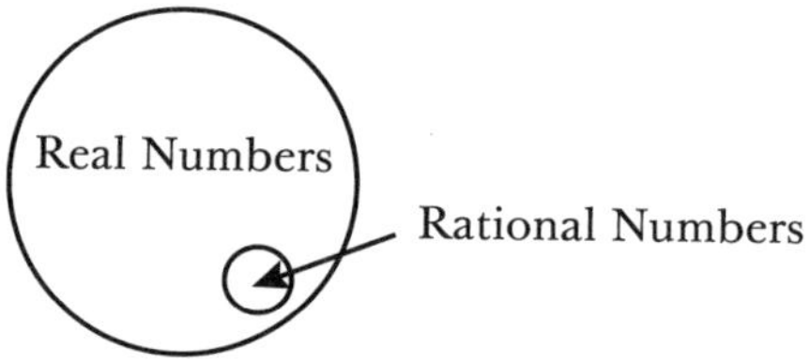

Real Numbers

Rational Numbers

In fact, the reality is more extreme than this picture. It turns out that there are *far* more real numbers than rational numbers. The infinity of the real numbers is so big that the infinity of the rational numbers looks like a speck in comparison. One way to understand the infinity that describes the set of real numbers is to return to our original discussion of elementary school "counting." Finite collections can be counted, plain and simple. You start at 1, and eventually you reach the end. Even large collections, say of one million items, may take a while to count, but eventually the counting does stop. The set of natural numbers is much larger than any finite set. It can be "counted," but the *counting never stops*. The same is true of the rational numbers. Using Cantor's creative diagonalization, we can point to each rational number and say, "first, second, third, fourth, …," guaranteeing that every rational number eventually gets named in the never-ending process. With the real numbers it is categorically different. It is not the case that the counting simply goes on and on forever. Rather, it is the case that there are so many that it can never get past "1." There are too many to list, even using an infinite list.

62 It can be shown that the "size" of the real numbers between 0 and 1 is the same as the "size" of the real numbers. One way of proving this is to use a scaled version of the tangent function to map numbers from one set to the other.

What *are* those numbers that are real but not rational? What is this set that makes up "the big part," or "most" of the real numbers? The answer is the *irrational numbers.*

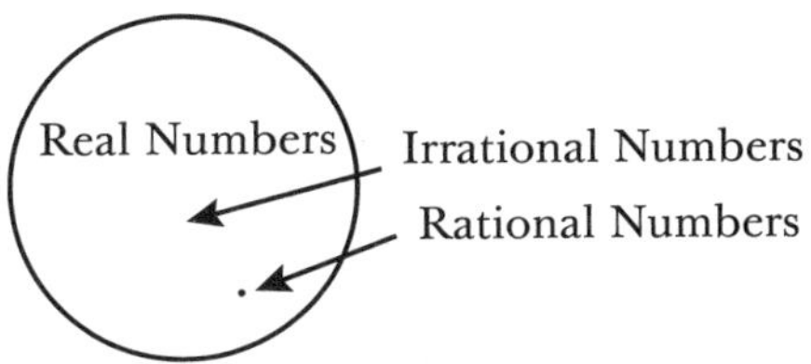

We therefore have three facts:

1. The size of the set of real numbers is a "big infinity."
2. The real numbers are made up of the rational numbers and the irrational numbers.
3. The size of the set of rational numbers is a "small infinity."

We can conclude that the irrational numbers are precisely those that make the size of the set of real numbers a "big infinity."[63] Moreover, we now have a proof that the irrational numbers are not countable, and this gives us a much larger set than the rational numbers.

> *The Uncountability of the Irrational Numbers Theorem*
> The irrational numbers are uncountable. Therefore, there are significantly more irrational numbers than rational numbers.

This is surprising because we went to great lengths earlier just to demonstrate that a single irrational number exists: $\sqrt{2}$. Moreover, most of us struggle to start naming irrational numbers: $\sqrt{2}$, π, some of the other square roots, and maybe combinations of these things, such as $\sqrt{2}+\pi$. Now we know that not only are there an infinite number of irrational numbers but *almost all of the real numbers are irrational.*[64]

63 A small detail should be worked out. We should convince ourselves that two small infinities cannot make a big infinity. In other words, maybe the rational and irrational numbers are both the "small" kind of infinity but when put together make a big infinity. But this is the very thing with which we dealt when arguing that the integers, which are really the "joining" of two countable sets (the non-negative integers and the negative integers), are countable. In other words, the union of two countable sets is always countable. That is to say, because the rational numbers are countable and the real numbers are uncountable, the irrational numbers *must* be uncountable. Moreover, Cantor's argument could be repeated verbatim with only the irrational numbers to show that they are uncountable.

64 There are so many irrational numbers that if you dropped a point randomly on a number line, the mathematical probability of that number being rational is 0 percent. This is counterintuitive, of course. How can an event be "possible" but have a probability of 0 percent? It is, however, amazingly true.

CHAPTER 10
THE SEVEN BRIDGES OF KÖNIGSBERG

Some problems, and their accompanying solutions, are important not merely for their general interest and the beauty displayed through a proof from *The Book*, but because they generate entirely new areas of mathematics. Such is the case for an eighteenth-century problem involving the East Prussian city of Königsberg, situated along the Pregel River.

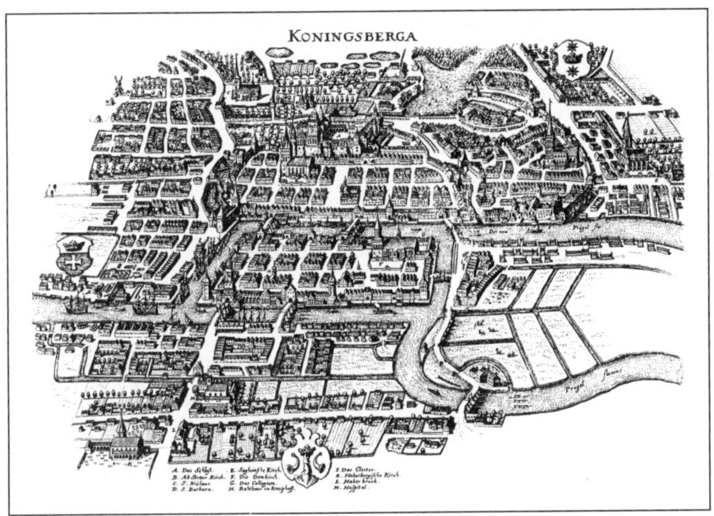

Königsberg in 1651[65]

The city, whose name literally meant "King's Mountain," was founded by Teutonic knights in 1254. In the middle of the city, the river splits into two branches, creating four distinct land regions. The central piece of land was the island of Kneiphof ("Pub Yard"), which served as a hub of city life.[66] To the east of Kneiphof (right on the map) was another island, appearing more like a peninsula on the map, named Lomse.[67] These two regions, together with the north and south banks, were connected

65 Merian-Erben, "A map of Königsberg from 1651." Available in the public domain: Retrieved from https://commons.wikimedia.org/wiki/File:Image-Koenigsberg,_Map_by_Merian-Erben_1652.jpg on September 16, 2024.

66 Hopkins, B. and Wilson, R., "The Truth about Königsberg." *The College Mathematical Journal*, volume 35, number 3 (May 2004). Retrieved from https://www.gss.ucsb.edu/sites/secure.lsit.ucsb.edu.germ.d7/files/sitefiles/news/conferences/euler/Hopkins1.pdf on September 16, 2024.

67 Königsberg has been since renamed as Kaliningrad and is located in Russia, with the Pregel River now named the Pregolya River. Kneiphof is now known as Kant Island in honor of the German philosopher Immanuel Kant, and Lomse is known as October Island.

by a series of seven bridges: Merchant Bridge, Blacksmith Bridge, High Bridge, Honey Bridge, Wooden Bridge, Köttel Bridge, and Green Bridge.

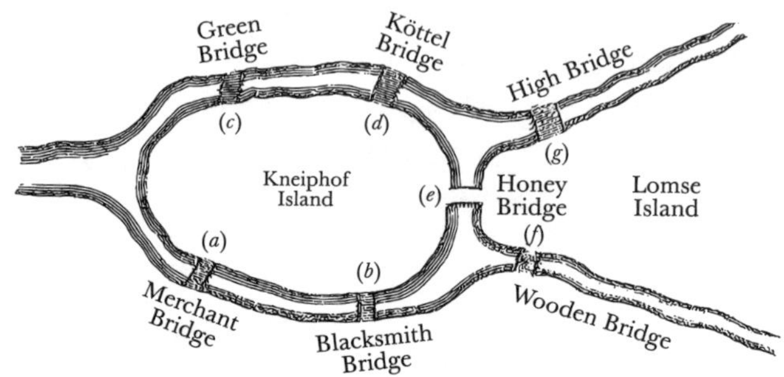

The Seven Bridges of Königsberg[68]

The people of Königsberg asked whether it might be possible to travel across each of the seven bridges exactly once. You may start wherever you like, and you may end wherever you like. For example, we might start on the island of Kneiphof. This seems like a good choice because from there we have the greatest number of options for bridges. We could then travel upward across Green Bridge (labeled *c*). From there, perhaps we cross High Bridge (*g*) onto Lomse, and Honey Bridge (*e*) back onto Kneiphof, Blacksmith Bridge (*b*), and Wooden Bridge (*f*). At this point, we are stranded on Lomse with no bridges to cross. After all, the only bridges off of Lomse are Wooden Bridge, Honey Bridge, and High Bridge, and we have already traversed each of those. In this attempt, we have crossed only five of the seven bridges: $c \to g \to e \to b \to f$, and have yet to visit Merchant Bridge (*a*) and Köttel Bridge (*d*).

Our mistake might have been the choice of starting location. What if we started on the bottom instead? Consider the path $a \to c \to d \to b \to f \to e$. We have done a bit better than our first attempt and in fact are quite close, having crossed all of the seven bridges except one. Unfortunately, we are now stranded on the island of Kneiphof with no access to the one bridge we still need to cross: High Bridge (*g*).

In fact, as much as we try to accomplish the challenge posed by the good people of Königsberg, we always seem to be at best one bridge short. We

68 Euler, L., "Figure 1." *Solutio problematis ad geometriam situs pertinentis*. Retrieved from https://commons.wikimedia.org/wiki/File:Solutio_problematis_ad_geometriam_situs_pertinentis,_Fig._1_-_Cleaned_Up.png on September 16, 2024. Labels added by the author.

are left wondering whether the task is impossible or merely quite difficult. To decide which is the case, we either need a valid solution or a proof of impossibility. For that, we turn to Leonard Euler, an eighteenth-century Swiss mathematician. Euler was actually much more than a mathematician. He made contributions to physics, astronomy, logic, engineering, and even geography. While it is unclear how he became aware of the Königsberg bridges question, we do know that he corresponded about the matter with Carl Leonhard Gottlieb Ehler, the mayor of Danzig,[69] a town in Prussia that was about fifty miles from Königsberg,[70] and eventually published a paper on the problem.[71]

In approaching this problem, Euler did something marvelously creative. Instead of focusing on the seven bridges, he focused on the four pieces of land.

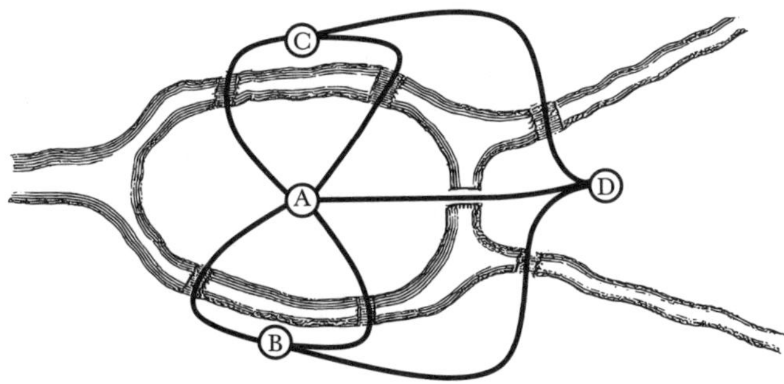

Following this change in perspective, Euler made a key observation that is somewhat similar to the informal law of gravity ("what goes up must come down"). In the case of bridges between lands, the rule becomes

69 The town of Danzig is now Gdansk, Poland.

70 Hopkins, B. and Wilson, R. "The Truth about Königsberg."

71 I have adopted both the bridge names (translated) and locations from Euler's work. It is, however, worth noting two things. First, the translation of the name for each bridge (*brücke* in German) will vary from source to source. Krämerbrückenfest (labeled Krämer by Euler) is known as Merchant Bridge or Shopkeeper Bridge. Schmiedebrücke (labeled Schmiede by Euler) is known as Blacksmith Bridge or Forge Bridge. Holzbrücke (labeled Holz by Euler) is known as Wooden Bridge. The Greunebrücke (labeled Grüne by Euler) is known as Green Bridge. Hohebrücke (labeled Hohe by Euler) is known as High Bridge. Honigbrücke (labeled Honig by Euler) is known as Honey Bridge or sometimes Honeymoon Bridge. It was apparently the only bridge that had an entry gate on it, and it led directly to the Königsberg Cathedral. Last comes Köttelbrücke, which is the most interesting. Euler labels this Köttel, which might be an archaic spelling of Kette, meaning chain or connection. For this reason, it is sometimes referred to as Connection Bridge. It may also refer to offal (the guts or giblets) of an animal. Therefore, the bridge is sometimes referred to as Giblet Bridge or Offal Bridge. Because of the ambiguity, I have retained the name of the bridge as Köttel. The second point of note is about Euler's diagram itself. For reasons unknown, the city has been reflected across a horizontal access. Therefore, in many pictures or maps of the city of Königsberg, the three bottom bridges are Green Bridge, Köttel Bridge, and High Bridge, whereas the bottom bridges are Merchant Bridge, Blacksmith Bridge, and Wooden Bridge. Euler's transformation is not merely a rotation that preserves the cardinal directions but rather a reflection. While it is true that Euler's map has North pointing downward, it is also true that Euler has Merchant Bridge to the east of Blacksmith Bridge. In reality, Merchant Bridge is west of Blacksmith Bridge.

"what comes in must go out." Let's consider the middle island of Kneiphof (*A*) and its five connections, or bridges, to other land masses: Blacksmith Bridge, Köttel Bridge, Green Bridge, Honey Bridge, and Merchant Bridge. If we cross onto Kneiphof, say using Blacksmith Bridge, we must leave Kneiphof using a different bridge, maybe Köttel Bridge. We continue our journey through other land masses, but eventually we must arrive at Kneiphof again, say using Green Bridge. Like before, we then depart using a different bridge, maybe Honey Bridge. At this point, we are now off the island of Kneiphof, but there is still one bridge remaining: Merchant Bridge. With an unpaired bridge left, we must conclude that it was either the first one traveled, in which case we started our journey on the island, or the last bridge traveled, in which case we ended our journey on the island. In other words, our journey must be in one of two categories.

Starting on Kneiphof

Merchant → … → Blacksmith → Köttel → … → Green → Honey → …

Ending on Kneiphof

… → Blacksmith → Köttel → … → Green → Honey → … → Merchant

What if we do the same kind of analysis with Lomse (*D*), which has only three connections (Honey Bridge, High Bridge, and Wooden Bridge)? Once on Lomse, say by way of Honey Bridge, we must leave Lomse, perhaps by way of High Bridge. The situation is the same as it was with Kneiphof. We have an unpaired bridge remaining: Wooden Bridge. Therefore, Lomse must also be either a starting place or an ending place of our journey. Putting these two things together means that a path crossing every bridge exactly once must either start on Kneiphof and end on Lomse or start on Lomse and end on Kneiphof.

The problem becomes stickier when we look at the other two regions to the north and the south: regions *B* and *C*. Like Lomse, each of these regions has three bridges. Therefore, both *B* and *C* also must be either starting points or ending points of our journey. This means that *all four* land masses have to be either starting points or ending points! Euler concludes, then, that the journey called for by the good people of Königsberg is not possible.

The problem occurred precisely because all four regions have an odd number of bridges connecting them to other regions.

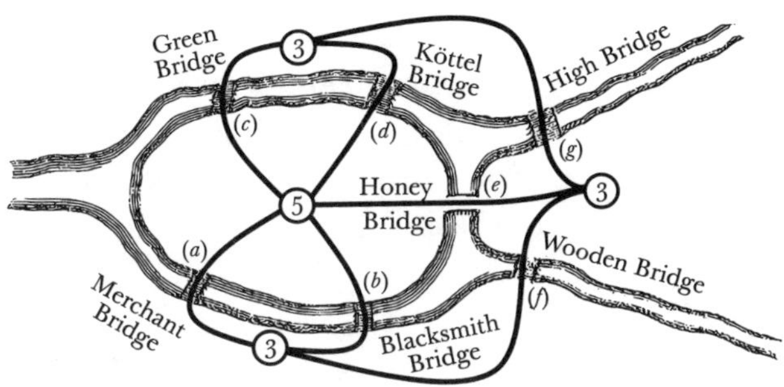

An odd number of bridges attached to a land mass does not itself cause a problem; it just tells us that such a location must be either the starting point or the ending point. The problem with Königsberg is that *there are simply too many odd-numbered lands*. More precisely, a bridge-traveling journey can handle only *two* odd-numbered lands: one for the starting location and one for the ending location.

What if we hire the city engineer, or maybe Euler himself, to construct an eighth bridge in the beloved city connecting the north area directly to the south area?

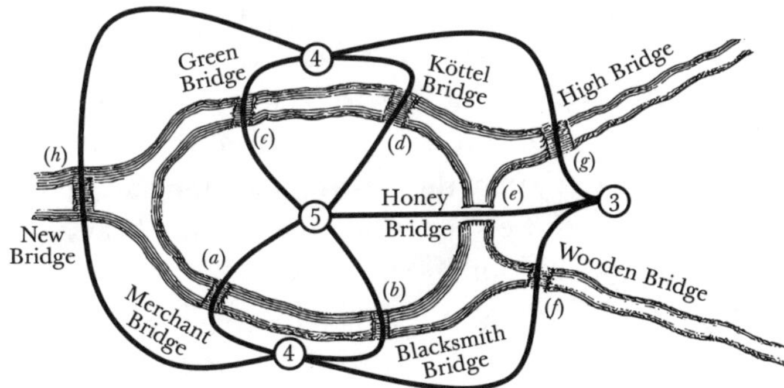

Now there are exactly two land masses with an odd number of bridges (Kneiphof and Lomse). The other two have an even number of bridges. We can conclude that Kneiphof and Lomse must be the starting and ending points of the journey and that there is hope for a bridge-crossing path.[72] For example, if we start our journey at Kneiphof, we can cross the bridges in this order: $a \to h \to c \to d \to g \to e \to b \to f$. Following this path carefully, we will see that the journey does indeed end on Lomse. This solution is not the only one, and in fact there are many solutions. In any of these solutions, though, the path will either start at Kneiphof and end at Lomse or start at Lomse and end at Kneiphof.

What about a different bridge configuration? What if the city engineer keeps New Bridge but demolishes the beautiful Honey Bridge?

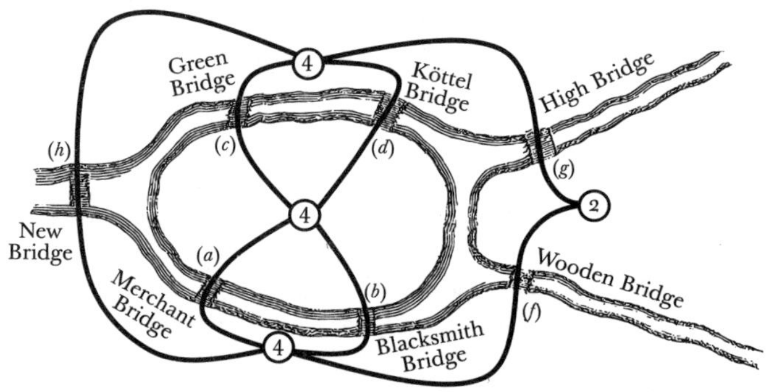

Now we have *no* odd-numbered lands. Does this mean that there is no starting point and no ending point? On the contrary, it means that the starting point *is* the ending point. Any path will have to end up back where it started. For example, starting again at the island of Kneiphof, a complete path across the bridges could be $a \to b \to d \to h \to f \to g \to c$. Trace this path carefully, and you'll find that the journey does indeed end back at Kneiphof. Try to change the starting location and find a different path. Regardless of where we start, the journey will always end back at the point from which we set out.

Euler generalized his findings in a 1736 paper.[73]

72 It should be noted that our argument does not show that a path is guaranteed, only that there is still some hope for one. Euler proved that in such cases a path is always possible, but I am not presenting that proof here.

73 Quoted in Hopkins, B. and Wilson, R., "The Truth about Königsberg." Euler's paper was written in 1736 but not published until 1741. Also, as noted above, while Euler presents a proof that a path is always possible if the number of land areas with an even number of bridges is either 0 or 2, we have only presented the proof that such a condition is necessary. In other words, we have only shown that if the number of areas with an odd number of bridges is something other than 0 or 2, a path is not possible. This is enough to answer the Königsberg question.

Generalized Bridges of Königsberg Theorem
If there are more than two areas to which an odd number of bridges lead, then such a journey is impossible. If, however, the number of bridges is odd for exactly two areas, then the journey is possible if it starts in either of these two areas. If, finally, there are no areas to which an odd number of bridges lead, then the required journey can be accomplished starting from any area.

Euler's simple question about bridges in the town of Königsberg eventually gave rise to a new area of mathematics called graph theory. In this context, a graph is a collection of vertices (akin to land areas in Königsberg) and edges connecting them (akin to the bridges of Königsberg). Through the creative move of focusing on land areas instead of bridges, Euler both discovered and revealed a structure in the problem that may not have been obvious at first. In fact, the revelation is so brilliant that one feels he *imposed* structure on the problem. This is not true, of course. The structure of a mathematical object, in this case a vertex-edge graph, is inherent in the object itself. It is up to the mathematician, through a genuine act of creativity, to *discover* that structure.

Much of mathematical discovery takes the same form: finding a veiled structure that is not at first evident but is key to unlocking the nature of the object and the solution to the problem. We saw a much simpler version of this when we dropped a vertical line through a triangle, thereby revealing symmetry, in the quest to find the triangle's area.

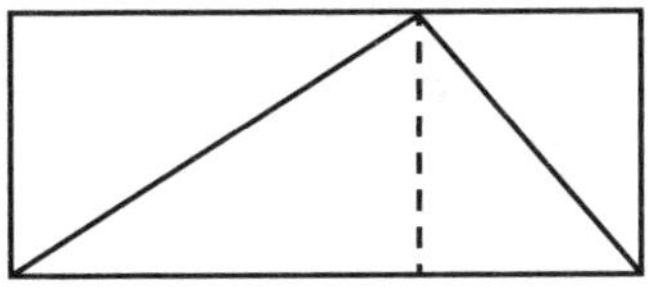

Another wonderful example of discovering a hidden structure happens in a well-known problem beginning with a 4×4 grid of squares after the top right and bottom left squares have been removed.[74]

74 The more well-known version deals with an 8×8 grid. The 4×4 version has the same solution but simplified graphics for our purposes.

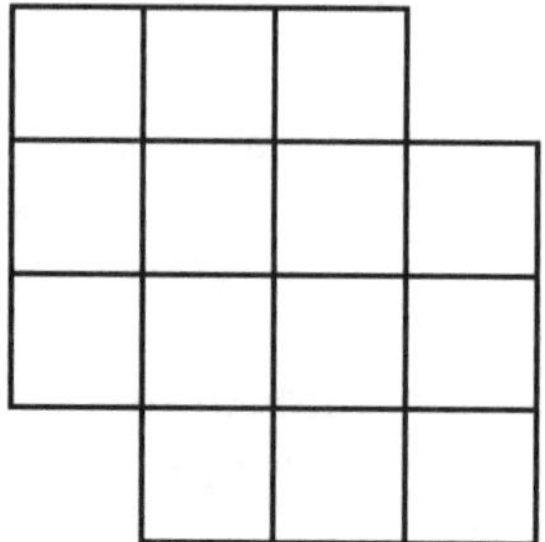

Is it possible to cover the entire shape with dominos (or 2×2 rectangles)? In theory, because there are an even number of squares, 14, this seems like it should be possible. Like the Königsberg Bridge problem, however, everything we try seems to end in frustration, with two squares left that are not next to each other.

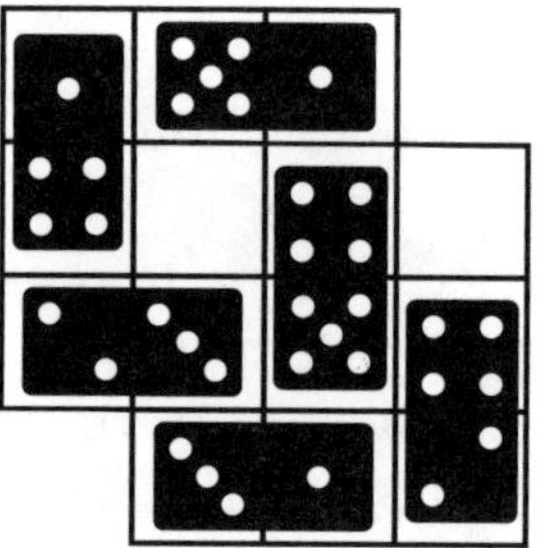

The structure waiting to be unveiled is a coloring of the grid itself using a checkerboard pattern.

Once this structure is brought out, the key observation is that every domino must cover one white square and one black square, and yet the number of white squares is greater than the number of black squares.

In other words, six dominos will cover all six black squares and six of the eight white squares. No matter what we do, there always will be two white squares left, as can be seen in our first attempt at covering.

There is a related problem about tracing paths inside a grid of squares, this time starting with a 5×5 layout. The question now is this: "Can we start at any point in the grid, move only left, right, up, or down, and visit each square exactly once?" For some starting locations, the solution is easy.

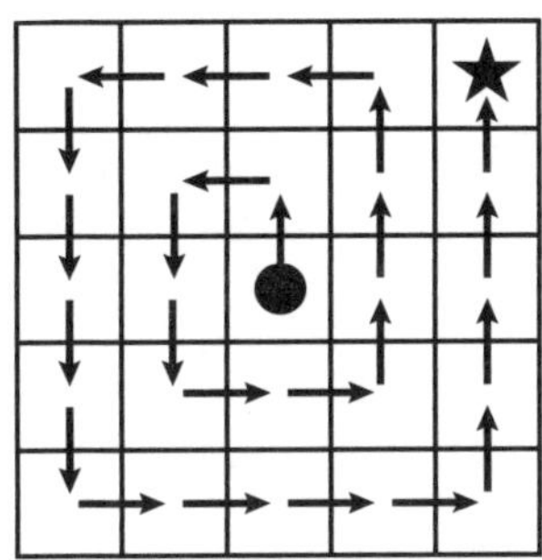

But some starting squares prove to be more difficult.

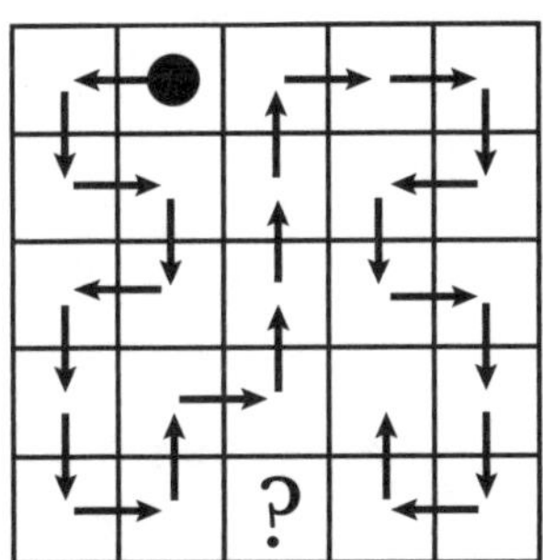

Are paths from such starting places simply difficult to find, or are they impossible, as was the case in the proposed Königsberg journey? We will not answer this question now, but perhaps finding a veiled structure in the object may help to provide both an answer to the question and the sketch of a proof.[75]

75 I have Dr. James Tanton to thank for the articulation of this problem in his video "What Made me a Mathematician" (https://youtu.be/b86QcxdrNF4).

PART III

TEN THEOREMS WORTH KNOWING, THOUGH THE PROOFS ARE DIFFICULT

CHAPTER 1
π IS AN IRRATIONAL NUMBER

In part i we proved that there are numbers that cannot be produced using the integers and the four basic operations of addition, subtraction, multiplication, and division. Specifically, we looked at the length of the hypotenuse in a 1×1 right triangle and argued that this number was irrational.

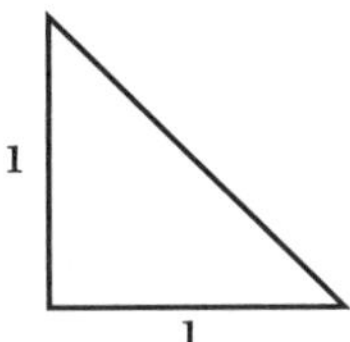

The first thing we said about this number, which is $\sqrt{2}$, was that it is a "reasonable" number. Only then was it interesting to know that the number cannot be written as a fraction of whole numbers. (Yes, this means that it is reasonable but not rational.) The first part is important: without it, our argument could be open to the objection that the number is not really a number at all. For example, the fact that the prime number in between 24 and 28 cannot be written as a fraction of whole numbers does not make it irrational because *there is no prime number between 24 and 28*. Perhaps in some way it is true that this nonexistent number cannot be written as a fraction of whole numbers, but it does nothing to show that irrational numbers exist.

The length of the hypotenuse is a perfectly reasonable number because the triangle itself is perfectly reasonable. The fact that its length, $\sqrt{2}$, is not rational confounded some ancient mathematicians precisely because the length is reasonable. We later saw, however, that this number is far from alone. There are *a lot* of irrational numbers. Even saying that there are an infinite number of irrational numbers does not do justice to just how many there are. There are far more irrational numbers than there are rational numbers. Nevertheless, $\sqrt{2}$ will always inspire some nostalgia in you and me as the first number whose irrationality we were able to prove. But it does not hold the prize of being the world's most famous irrational number. Those bragging rights undoubtedly go to the number π.

Mathematics students all over the world have asked their teachers, "How do we know that π is irrational?" It is an understandable question. There are, after all, some *really* good approximations for π. For example, the number $\frac{1{,}146{,}408}{364{,}913}$ is equal to 3.1415926535914..., which matches π to eleven decimal places. How do we know that fractions like this will only ever approximate and will never equal π? While many an inquisitive student may have asked this, it is a rare student who wonders about the more fundamental question, "How do we know that π is even a number?"

What? How do we know π is a number? How could it not be? To appreciate the reasonableness of the question, we need to first define π. Note that it does no good to define it generally as "a number with decimals that do not repeat and that never end." This is a property of all irrational numbers. Nor does it help to say, "π is 3.14159, and so on." There are infinitely many numbers that start off 3.14159. Defining the number $\sqrt{2}$ was a bit easier. Geometrically it is "the length of the hypotenuse in a right triangle with both legs equal to 1." Algebraically, it is "the number that when multiplied by itself equals 2." *But what about π?* To define this giant of a number, we begin by looking at one of the most basic geometric shapes: the humble circle.

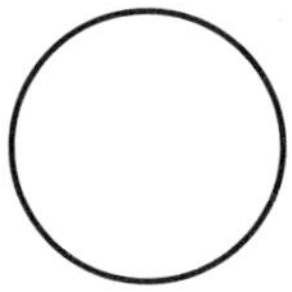

We start by measuring the distance across the circle, a length known as the *diameter*.

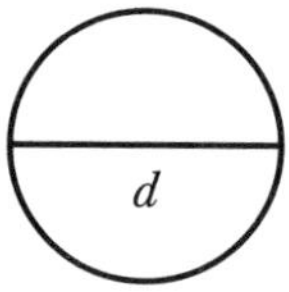

d

Then we "unroll" the circle into a line and measure the length of that line, which represents the distance around the circle, or the *circumference*.

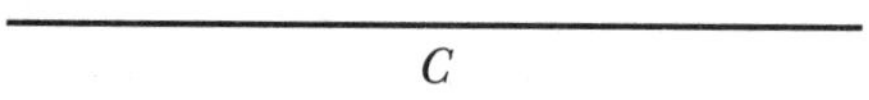

C

Finally, we divide the circumference by the diameter: $\frac{C}{d}$. The result is a number, which we name π. This process is the very *definition* of π.

DEFINITION: The number π is the ratio of the circumference to the diameter in any circle.

The problem with this definition is at the heart of the question we asked earlier: "How do we know that π is a number at all?" Maybe a more precise question is "How do we know that π is only *one* number?" After all, the number was defined by a *process*, and the process started by drawing a circle—*any* circle. How do we know that one hundred people, drawing one hundred circles, measuring one hundred diameters and circumferences, and dividing those two values *will always get the same result?* If they don't get the same result, then our definition makes no sense. If we get multiple values, then this is as good as the number's not existing. If the number does not exist, then the question of whether or not it is rational is nonsensical.

How *do* we know that the ratio of the circumference to the diameter is always the same?[76] What I offer here is not a formal proof but rather a pretty good intuitive argument. We have all seen scale models from time to time. Some scale models exist as actual models, as when a replica of a Lamborghini is made at one-twelfth its original size. Other scale models exist on paper, as when one inch on a map represents one mile in reality. Regardless of where we have seen them, scale models always have one thing in common: all lengths are increased or decreased by the same factor ("proportionally") when moving back and forth between a model and the real thing.

For example, suppose that the rectangle on the left is a scale model of the one on the right.

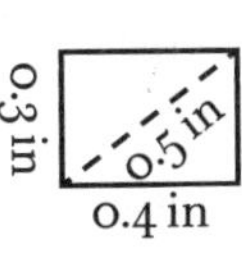

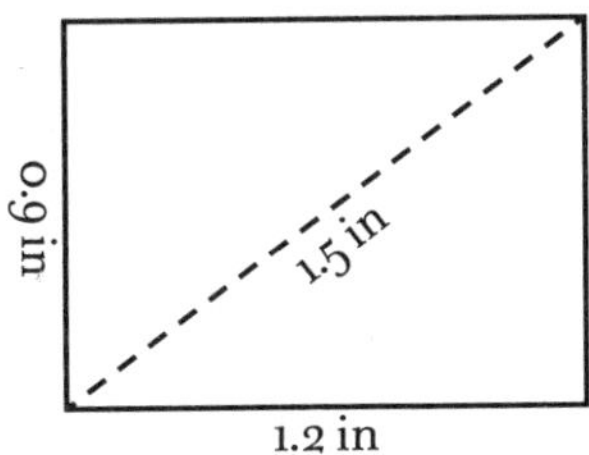

76 Note that it is not good enough to say, "I remember that the formula for the circumference of a circle is $C = \pi d$, so dividing both sides by d gives me $\frac{C}{d} = \pi$. Therefore, the ratio of the circumference to the diameter is constant, and in fact is π." This reasoning is circular (in both senses of the word). The formula for circumference is based on π's being a constant to begin with, so it cannot be used to justify that π is a constant.

The *scale factor* is 3. In other words, in terms of the side lengths, the larger rectangle is three times as big as the smaller rectangle. Equivalently, the smaller rectangle is one-third the size of the larger rectangle. Note that this relationship of measurements is true not only for the sides but also for the diagonal. The smaller rectangle has a diagonal of 0.5, which is one-third of 1.5, the length of the diagonal of the larger rectangle. In fact, this same relationship holds for any measurement of lines, including the perimeter. The perimeter of the smaller rectangle is $0.3+0.4+0.3+0.4=1.4$, and the perimeter of the larger rectangle is $0.9+1.2+0.9+1.2=4.2$, which is three times 1.4.

This same thing does not work for area. In that case there is a different but equally predictable relationship. The area of the smaller rectangle is $0.3\times 0.4=0.12$ in^2, and the area of the larger rectangle is $0.9\times 1.2=1.08$ in^2. The value 1.08 is nine times 0.12. It is not an accident that 9 is the same as 3^2. I will leave it to you to figure out the general rule, and perhaps even to extend it to the relationship of volume if we are dealing with three-dimensional models.

At this point we have a lemma.

> *The Scale Model Lemma*
> Whenever one shape is a scale model of another shape, the scale factor applies to any length in the object, including sides, diagonals, and perimeters.

We can restate this lemma in perhaps a more useful way for our discussion of π. If we take the ratio of any two lengths in the model, it will be the same as the ratio of the corresponding lengths in the real thing. For example, if we calculate the ratio of the width to the length in the smaller rectangle, we get $\frac{0.3}{0.4}=0.75$. If we calculate the ratio of the width and length in the larger rectangle, we get $\frac{0.9}{1.2}$, which is also 0.75. This same ratio will be found using any pairs involving corresponding sides, diagonals, or even the perimeters. For example, the ratio of the perimeter to the diagonal in the smaller rectangle is $\frac{1.4}{0.5}=2.8$, and the ratio of the perimeter to the diagonal in the larger rectangle is also $\frac{4.2}{1.5}=2.8$. This is not a mystery, actually. It follows from having a common scale factor. We already noted that because 3 is the scale factor, we must have $4.2=1.4\times 3$ and $1.5=0.5\times 3$. Therefore,

$$\frac{4.2}{1.5} = \frac{1.4 \times 3}{0.5 \times 3} = \frac{1.4}{0.5}.$$

> *The (Restated) Scale Model Lemma*
> Whenever one shape is a scale model of another shape, the ratio of two lengths in the model is equal to the ratio of the corresponding lengths in the original object.

Additionally, I propose a second lemma.

> *The Scale Models of Circles Lemma*
> All circles are scale models of one another.

I will not formally prove this lemma. But drawing a bunch of circles will give us a pretty good sense that it is true. Circles are so basic, so symmetric, that they are all scale models of one another.

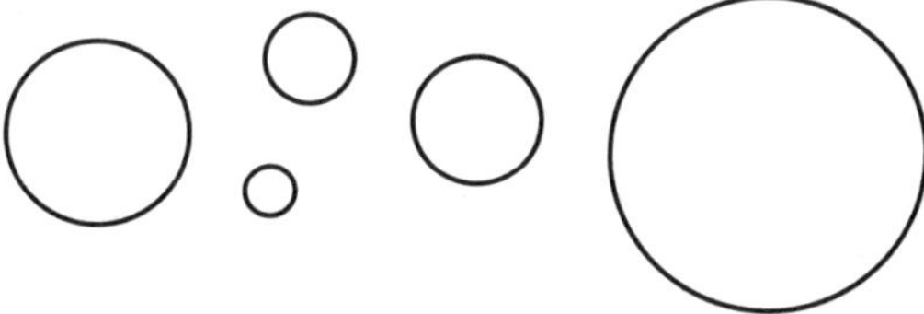

The same thing is true of regular polygons, by the way. All equilateral triangles are scale models of one another, all squares are scale models of one another, etc.

Because all circles are scale models of one another, the ratio of any two lengths must be the same. Specifically, this means that the ratio of the perimeter (circumference) to the diagonal (diameter) is always the same, no matter the circle. We call this ratio π. This whole thing may seem tedious, but it was a very important step in showing that π is irrational. We can at least take comfort in knowing that π exists.

The π-is-Actually-a-Number Theorem
The number π, which is the ratio of the circumference to the diameter in any circle, is an actual number because this ratio is always the same, no matter the circle.

We can now worry about whether this number is rational. The truth is that it is not, but that is a difficult thing to prove. But the irrationality of π is absolutely worthy of including in this book's list of theorems everyone should know, even if arguing it is hard.

Irrationality of π Theorem
The number π cannot be written as the ratio of two whole numbers and hence is irrational.

The mathematician Johann Heinrich Lambert first proved this theorem in 1768. As an irrational number, its decimal expansion never terminates and never repeats. In fact, the digits that are in the number π are generally thought to be *normal*. This means that any digit or sequence of digits is equally likely to show up. Therefore, the digit 4 shows up as often as 7, and the sequence 285 shows up as often as 123.[77] In some ways, it also means that the digits of π are "without pattern,"[78] which is why they are used by computers for generating "random" numbers. This lack of pattern is not true of all irrational numbers. We have already seen that the number 0.1010010001000010000010000001... is irrational, but its digits are certainly not normal. The digit 0 shows up far more often than the digit 1 does, and the digit 7 does not show up at all. Moreover, the digits also have a clear pattern.

Lest we think that π itself is inherently without pattern, there are some rather beautiful expansions of it that have been proven throughout the centuries. In 1665, the English mathematician John Wallis found a wonderful formula.

$$\frac{\pi}{2} = \left(\frac{2}{1}\times\frac{2}{3}\right)\times\left(\frac{4}{3}\times\frac{4}{5}\right)\times\left(\frac{6}{5}\times\frac{6}{7}\right)\times\left(\frac{8}{7}\times\frac{8}{9}\right)\times\cdots$$

77 The normalcy of π's digits has neither been proven nor disproven.

78 I use the phrase "without pattern" hesitantly. If we were to start reciting these "random numbers" as "3, 1, 4, 1, 5, 9, ...," many a mathematician would say, "Those are the digits of π!" rather than "Those numbers seem random to me." In other words, because π is so famous, one could argue that the sequence of digits is a pattern that is a result of its own fame.

The numerators in each pair are the even numbers, and the denominators in each pair are successive odd numbers. If all infinite terms are multiplied together, the result is $\frac{\pi}{2}$. If we want this to be written as a formula for π, we can multiply both sides of the equation by 2.

$$\pi = 2\times\left(\frac{2}{1}\times\frac{2}{3}\right)\times\left(\frac{4}{3}\times\frac{4}{5}\right)\times\left(\frac{6}{5}\times\frac{6}{7}\right)\times\left(\frac{8}{7}\times\frac{8}{9}\right)\times\ldots$$

We should not be bothered by the infinite number of terms. By this point, we have dealt with various ways of combining an infinite number of terms to obtain a finite result.

Around the same time, the German mathematician Gottfried Leibniz proved that

$$\frac{\pi}{4} = 1-\frac{1}{3}+\frac{1}{5}-\frac{1}{7}+\frac{1}{9}-\cdots,$$

or

$$\pi = \frac{4}{1}-\frac{4}{3}+\frac{4}{5}-\frac{4}{7}+\frac{4}{9}-\cdots.$$

Earlier, in the mid-1400s, Nilakantha Somayaji discovered another formula for π.

$$\pi = 3+\frac{4}{2\times3\times4}-\frac{4}{4\times5\times6}+\frac{4}{6\times7\times8}-\frac{4}{8\times9\times10}+\cdots$$

Finally, Leonard Euler solved the eighty-year-old Basel problem in proving an extremely elegant formula.

$$\frac{\pi^2}{6} = \frac{1}{1^2}+\frac{1}{2^2}+\frac{1}{3^2}+\frac{1}{4^2}+\frac{1}{5^2}+\cdots$$

We close this chapter with three infinitely embedded fractions, or what mathematicians call *continued fractions*, that also equal π.[79]

79 We will not go into the formal definition of a continued fraction here. Of note, though, Lambert used a continued fraction for the function $\tan(x)$ in his proof for the irrationality of π.

$$\pi = \cfrac{4}{1 + \cfrac{1^2}{2 + \cfrac{3^2}{2 + \cfrac{5^2}{2 + \cfrac{7^2}{2 + \cfrac{9^2}{2 + \cdots}}}}}}$$

$$\pi = \cfrac{4}{1 + \cfrac{1^2}{3 + \cfrac{2^2}{5 + \cfrac{3^2}{7 + \cfrac{4^2}{9 + \cfrac{5^2}{11 + \cdots}}}}}}$$

$$\pi = 3 + \cfrac{1^2}{6 + \cfrac{3^2}{6 + \cfrac{5^2}{6 + \cfrac{7^2}{6 + \cfrac{9^2}{6 + \cfrac{11^2}{6 + \cdots}}}}}}$$

Any of these infinite sums, products, or continued fractions can be used to find simple fractions that are approximations of π.

$$\pi \approx 3 + \frac{4}{2\times3\times4} - \frac{4}{4\times5\times6} + \frac{4}{6\times7\times8} - \frac{4}{8\times9\times10} = \frac{989}{315}$$

All of this is to say that even though the digits of the world's most famous irrational number do not seem to exhibit any pattern, this does not mean that beautiful patterns are absent in other representations of π.

CHAPTER 2
THE AREA OF A CIRCLE

In the first part of this book, we discussed why the formula for the area of a triangle works. Our strategy was to draw a picture showing that a triangle is half of a rectangle. Therefore, if the area of a rectangle is "length times width," or $l \times w$, then the area of the triangle must be $\frac{1}{2} \times \text{length} \times \text{width}$, or $\frac{1}{2} \times l \times w$. More commonly, we write $A = \frac{1}{2} \times b \times h$.

Some areas can be calculated using a bit more "surgery." Consider finding the area of a parallelogram.

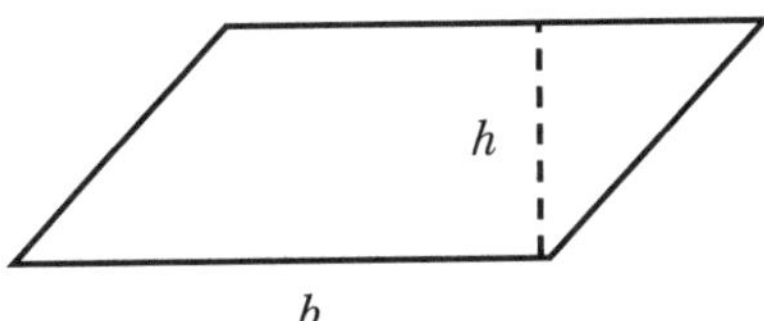

Here, we can cut off the triangle that is formed by the height and slide it over the left. The shape is now a rectangle, so the area is $b \times h$.

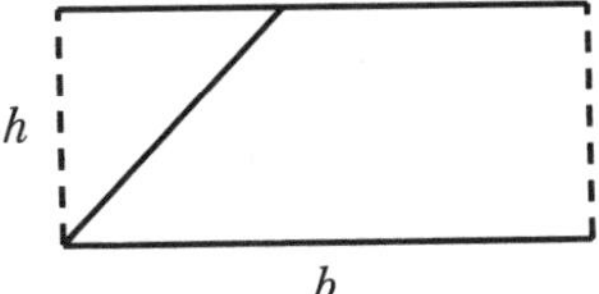

If we were being rigorous, there is more to prove here, specifically how we know that the shape actually is a rectangle, i.e., that it has four right angles. I leave that problem for you to work out, but for now we are satisfied with this argument.[80]

For other, more complicated polygons, we can often split them into familiar shapes with known area formulas. In fact, we saw in our work on polygonal tiling patterns that every polygon can be split into triangles. As a last resort, we could follow that route. That said, our task in this chapter deals with a shape that is very much *not* a polygon. We are interested in the area of a circle. In the previous chapter we determined that

80 There is another nuance that is much harder to deal with. If the parallelogram is so "sheared" that the entire top is completely to the right of the bottom, the triangle cutting does not work. This does not hurt the main argument of this chapter, but it is an interesting problem to think about. How would you deal with a "very sheared" parallelogram?

the perimeter of a circle, known as the circumference, was $C=\pi\times d$, or "π times the diameter." To be precise, what we really discovered was that the ratio of the perimeter to the diameter was constant, and we named that constant π. In other words, the formula for circumference comes directly from the definition of π.

One challenge with calculating the area of a circle is that it cannot be broken up into triangles. In fact, it cannot be broken up into any polygons, because a finite number of straight lines can never be combined to make a curve. This is partly why the Greeks consider the circle and the line to be the two basic shapes in the universe, leading to the two basic tools of the geometer, the compass and the straightedge. The circle and the line are decidedly *different things*.

While a formal proof for the area of a circle is difficult, we can still give a very compelling argument for it by slicing up the circle and rearranging the pieces. The way we slice the circle will be exactly how we slice a pie. (The spelling is intentional; as much as this discussion is about π, I really do mean a *pie*.)

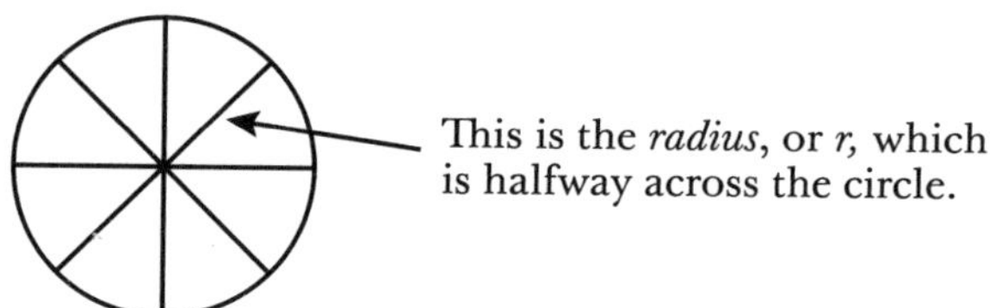

When we slice up the circle, we can rearrange the pieces in an alternating pattern.

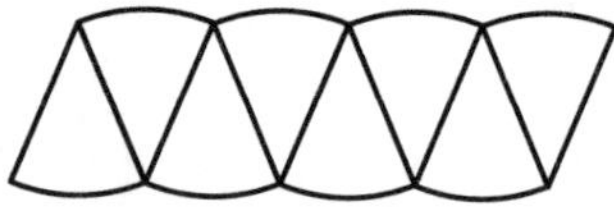

The reconstituted shape looks a lot like a parallelogram. To be sure, it is *not* a parallelogram because it has curves on the top and the bottom. That said, just for a minute let's *pretend* that this shape is a parallelogram. What would be the height and the base? The height is easier to see because it is a radius of the circle from where these slices came. What about the base? The length of the base is formed by the "crust" of the pie slices. Because we used all of the slices, the entire crust is present in our pseudo-parallelogram, with half on top and half on the bottom. Therefore, the base is "half the crust," and because the "crust" is the circumference (perimeter) of the circle, the base is exactly half the circumference. We noted at the start of this chapter that the circumference is $\pi \times d$, but the radius is, by definition, half the diameter. Therefore, if the entire circumference is $\pi \times d$, then half the circumference is $\pi \times r$.

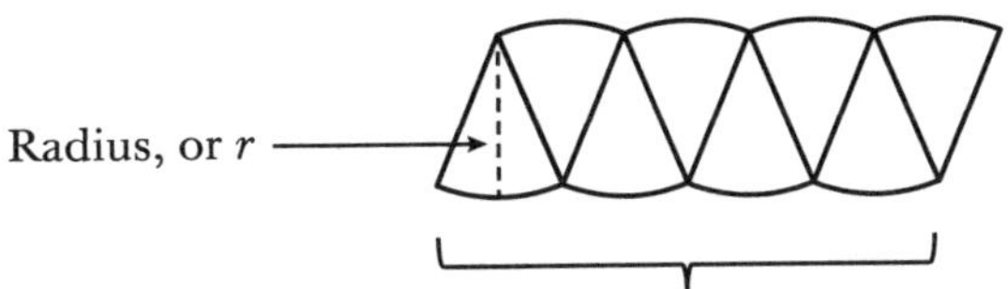

We are now ready to find the area of our circle by pretending that the pseudo-parallelogram really is a parallelogram. The area of a parallelogram is "base × height," and so the area of our "parallelogram" is

$$A = (\pi \times r) \times r,$$

or

$$A = \pi \times r^2.$$

You would be right to object that the shape was not really a parallelogram, and so we have no right to use its associated area formula. But what happens if we increase the number of slices to *sixteen*?

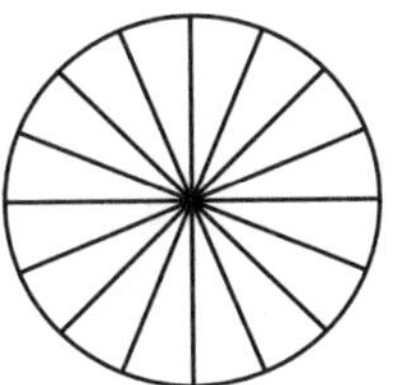

When we rearrange the slices, we still do not get a perfect parallelogram, but it is much closer.

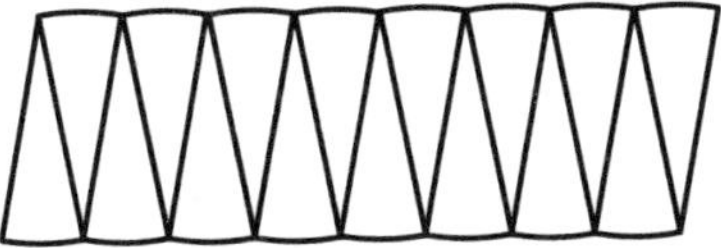

The height is still the radius (r), and the base is still half the circumference ($\pi \times r$). Therefore, the area is still, sort of, $\pi \times r^2$. I say "sort of" because the pseudo-parallelogram is *still* not actually a parallelogram, however close it may be. But what if we didn't use eight slices or sixteen slices, but instead used 1,000,000 slices? I will not draw the picture here, but if we did, three things would be certain: the height would still be r, the base would still be $\pi \times r$, and you would not be able to visually distinguish it from a parallelogram.

Does it matter that we cannot distinguish the shape from a parallelogram and yet know that it is decidedly *not* a parallelogram? For the mathematician, it definitely matters. That is precisely why this argument is in this section on "Theorems Worth Knowing, Though the Proofs Are Difficult" and not the section on "Proofs Worth Knowing." What I have presented here is *very good evidence* that the area of a circle is $\pi \times r^2$, but it is *not a proof*.

Mathematicians have worked on this problem for as long as they have known about circles. Another argument uses either *inscribed* or *circumscribed* regular polygons. The diagram on the left is a circle that has an inscribed regular octagon, and the one on the right has a circumscribed regular octagon.

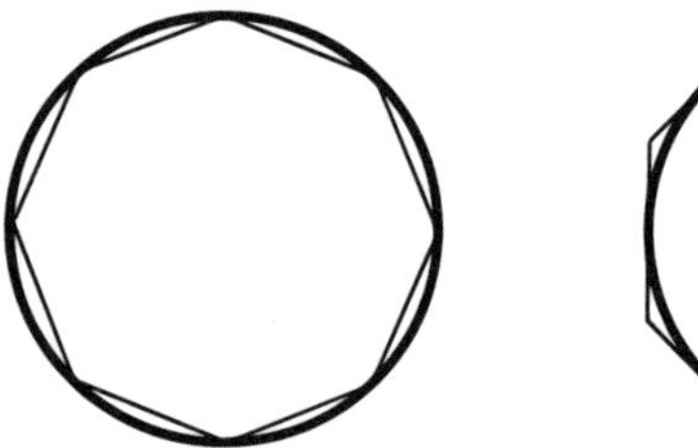

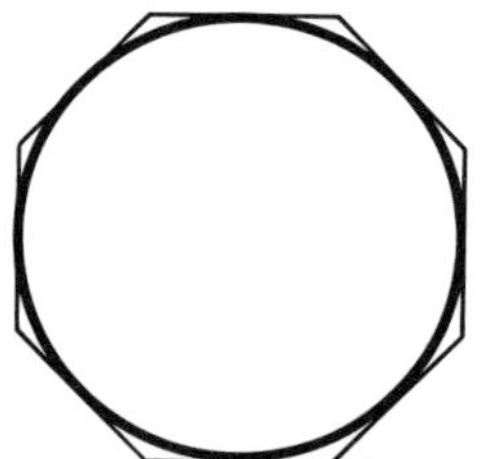

The argument using the circumscribed version begins by slicing the octagon into eight triangles.

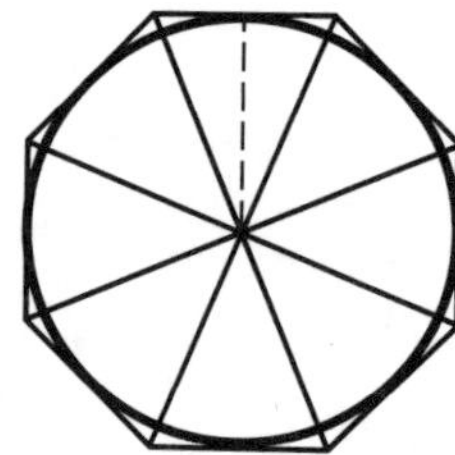

The dotted line is called an *apothem*. It is the line that starts from the center of a regular polygon and intersects a side at its midpoint. The intersection also happens to be at a right angle, making it the height of the triangle as shown in the diagram. The apothem serves as a sort of "radius" for the polygon, and for our purposes it also happens to be the radius of the circle. It is helpful to pull off one of the triangles.

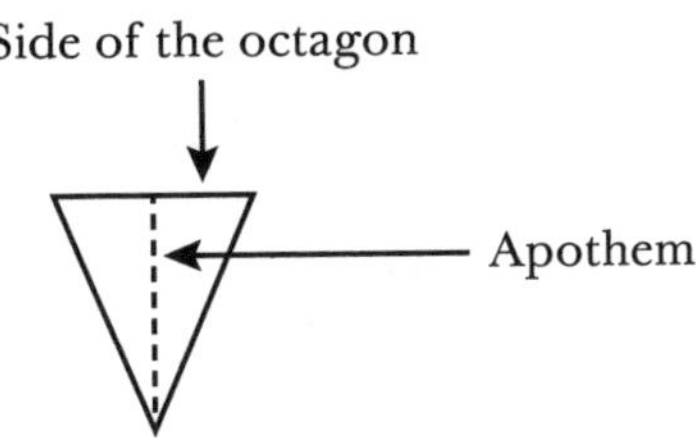

The area of this triangle is $\frac{1}{2} \times$ octagon side $\times$ apothem. There are eight triangles around the octagon, so the total area of the octagon is $8 \times \frac{1}{2} \times$ octagon side $\times$ apothem. We can rearrange the terms to group the 8 and the length of the octagon side.

$$A = \frac{1}{2} \times (8 \times \text{octagon side}) \times \text{apothem}$$

"$8 \times$ octagon side" is precisely the perimeter of the octagon because there are eight "octagon sides" around an octagon. Therefore, we have a very nice formula for the area of the octagon.

$$A = \frac{1}{2} \times \text{perimeter} \times \text{apothem}$$

This equation holds generally: it does not matter if it is a regular octagon, a decagon, or a 1,000,000-gon. The area will always be $\frac{1}{2} \times$ perimeter $\times$ apothem. I leave it to you to convince yourself that this is the case no matter how many sides there may be.

Even with eight sides, we can see that the circle is *almost* the same as the regular octagon. If we instead used a regular 1,000,000-gon, we would not be able to see the difference between the polygon and the circle. If the area of the regular polygon is $\frac{1}{2} \times$ perimeter $\times$ apothem, and if for large values the circle is nearly indistinguishable from the polygon, then the circle's area must be very near to $\frac{1}{2} \times$ perimeter $\times$ apothem. The circle's "perimeter" is its circumference, $\pi \times d$, and we have already seen that half the circumference is $\pi \times r$. Further, its "apothem" is its radius (r). Thus, the area is $(\pi \times r) \times r$, or $\pi \times r^2$.

The Greek mathematician Archimedes gave a similar proof using his famous "method of exhaustion," which is one of the first attempts at defining a *limit*, the fundamental concept of calculus. In fact, it is modern calculus that we need in order to write a formal and rigorous proof for the area of a circle. Nevertheless, the argument by slicing and rearranging into a parallelogram and the argument by circumscribing the circle with a regular polygon are very convincing and certainly worth knowing.

Area of a Circle
The area of a circle with radius r is $\pi \times r^2$.

CHAPTER 3
THE INDEPENDENCE OF THE PARALLEL POSTULATE

A WELL-KNOWN RIDDLE goes something like this:

> *A bear travels one mile directly south, then turns and travels a second mile directly east, and finally turns again for a third mile heading directly north, upon which the bear ends exactly where he started. What color is the bear?*

The solution to the riddle has everything to do with geometry on the surface of a sphere, in this case the Earth. We like to think of movement as if it is on a flat plane so that successive journeys of south, east, and north would create an "open square," making a return to the starting point impossible without a fourth leg heading west. In this way, our intuitive understanding of geometry is very much how Euclid conceived of it.

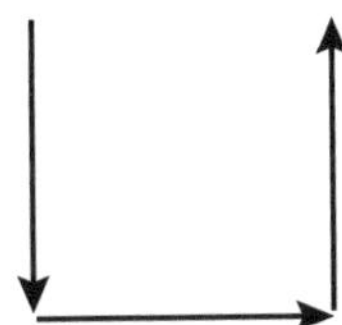

Geometry on a sphere does not work this way, and with some thought, we can resolve the apparent paradox in the riddle, which then might help us to determine the color of the bear. We will return to the riddle in a bit, but for now the important thing is that different surfaces have different geometries. The existence of these different geometries has everything to do with one of the most important problems in the history of mathematics.

All mathematical systems have sets of starting points from which the other truths are derived. Some of these starting points are *definitions*, but these exist more to provide a common language rather than to build a solid foundation for the eventual edifice of theorems. The solid foundation for Euclid is a set of five *postulates*, which he took to be "self-evident statements" that, by their nature, do not require proof.[81] The entirety of the *Elements* is built upon these five statements.

1. To draw a straight line from any point to any point.
2. To produce a finite straight line continuously in a straight line.
3. To describe a circle with any center and distance.
4. That all right angles are equal to one another.
5. That, if a straight line falling on two straight lines makes the interior angles on the same side less than two right angles, the two straight lines, if produced indefinitely, meet on that side on which are the angles less than the two right angles.

The first three postulates claim that things can be accomplished. They say, in short, that finite lines can be drawn by joining two points, lines can be extended into longer lines of any length, and circles of any radius and center can be drawn. Philosophically, they establish the existence of the two basic objects in geometry: lines and circles.

The fourth postulate might seem obvious at first, claiming that all right angles are equal (or, in modern parlance, *congruent*). After all, if two angles each measure 90°, are they not clearly equal? The issue here is that a right angle is not *defined* in terms of degrees. For Euclid, a right angle is the angle formed when a line is set up on another line forming two equal angles.

If these two angles are equal,
then they are both right angles.

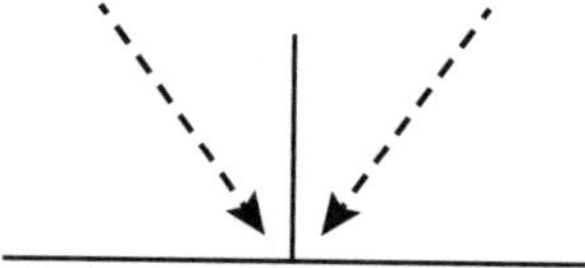

The question now is "If we create two copies of this situation, will the right angles from each copy be equal?"

81 In antiquity both *postulate* and *axiom* were used for a self-evident statement, but postulates were specific to the field of study (in this case geometry), whereas axioms were common to several fields. Euclid opts for the term *common notions* instead of axioms for these more general principles that extend beyond geometry. Euclid includes five common notions: (1) Things which are equal to the same thing are also equal to one another; (2) If equals be added to equals, the wholes are equal; (3) If equals be subtracted from equals, the remainders are equal; (4) Things which coincide with one another are equal to one another; and (5) The whole is greater than the part. These apply to shapes, but they also apply to numbers, and as such are more general common notions rather than geometry-specific postulates.

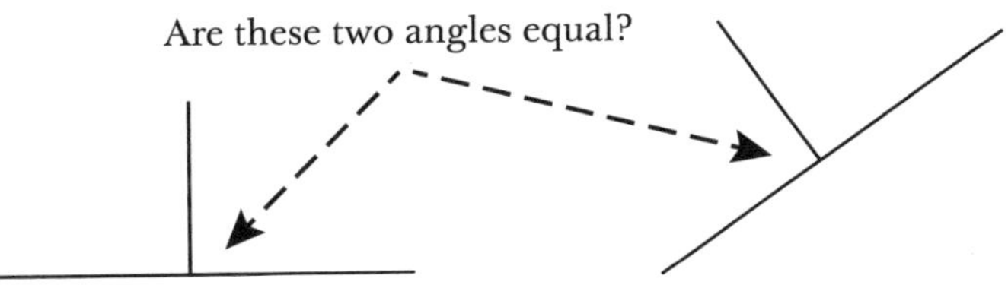

The fourth postulate says yes. Throughout the *Elements*, the fourth postulate becomes the basis for comparing angles and claiming that two angles are equal.[82]

The fifth postulate is entirely different. Even a basic word count sets it apart—it is longer than the other four combined—but that is not the only thing that stands out about Postulate 5. The other four seem easy enough to understand and easy enough to believe as "self-evident." The last postulate is not so easy to understand and even causes us to pause and think about whether or not it is true. A diagram can at least help us understand *what* it says. Euclid starts with "a straight line falling on two straight lines makes the interior angles on the same side less than two right angles." The phrase "falling on" means that a line crosses two other lines. The phrase "less than two right angles" requires a bit more parsing. As we saw in the chapter on the Triangle Angle Sum Theorem, Euclid does not use degrees to measure angles; his only unit of measure for angles is "one right angle." In fact, we also saw in that same chapter that "two right angles" really means a straight line, or 180°. Therefore, this first phrase means, "A line crosses two other lines so that the interior angles on one side of that line add to less than 180°."

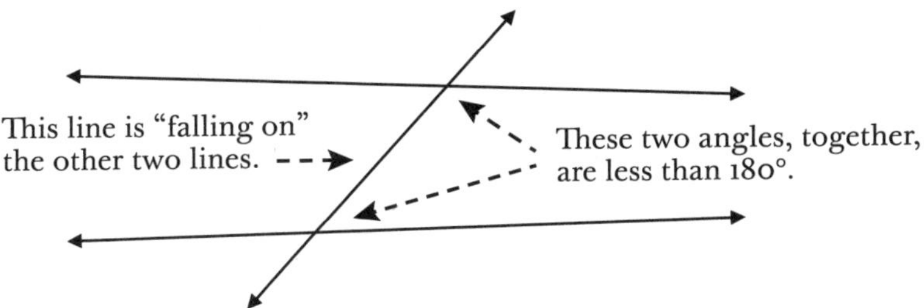

82 Equal segments are established by the ability to draw circles and the definition of a circle. All radii of a circle are equal by definition.

The "conclusion" of the fifth postulate is that when the two lines are "produced indefinitely," they will eventually meet on the side that has the two aforementioned angles. In other words, the lines are not parallel.

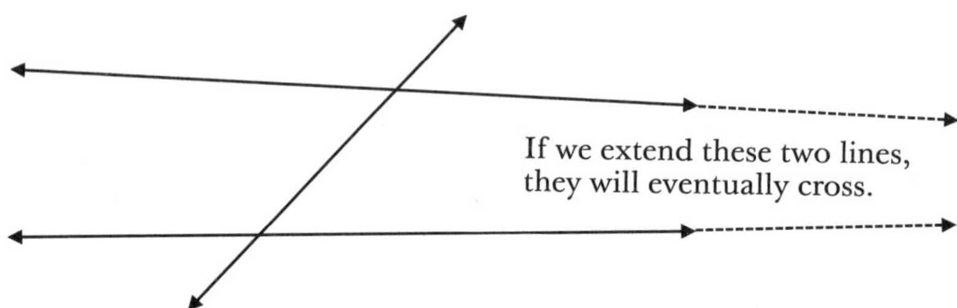

The picture makes it somewhat convincing that the postulate is true, but is it "self-evident"? Can we really just state it without proof and have the same level of certitude as something like "two points can be joined with a line" or "a circle can be drawn"? If you have even a small doubt about the self-evident nature of this postulate, you are not alone. In fact, for years it was thought that this postulate could be proven with the other four. If such a proof exists, it should not be listed as a postulate but rather as a theorem (or for Euclid a *proposition*). Yet time and again, the proof remained elusive as many a mathematician tried unsuccessfully to find it.

There are several reasons why mathematicians would be interested in finding a proof. One is the simple desire to minimize the number of assumptions that form the foundation of any mathematical system. A second is its importance in Euclid's geometry. The third, and most obvious, is that a proof would confirm that the property is *true*. After all, the Parallel Postulate is the very foundation of everything Euclid has to say about parallel lines.[83] For this reason, it became known as the Parallel Postulate. It appears for the first time in Euclid's proof of Proposition 29, which is what we called the Z Lemma in our warm-up,[84] and the Z Lemma was used to prove the Triangle Angle Sum Theorem. In other words, the

83 It is interesting that Euclid chooses to base his theory of parallel lines on a postulate that specifically deals with lines that are not parallel. He could have given a similar statement in positive terms: if two parallel lines are crossed by a third line, then the two interior angles on the same side of the crossing line are equal to two right angles. The actual Parallel Postulate includes a bit more information (which side the lines cross on), but still, Euclid must have thought that the version he gave was "more self-evident" than an equivalent version of it based on parallel lines.

84 Proposition 29 includes the congruence of alternate interior angles when parallel lines are cut by a transversal (the Z Lemma), but it also includes the congruence of what modern geometry texts call corresponding angles, as well as the supplementary nature of same-side interior angles.

fact, or at least the proof, that the angles in a triangle add to a straight line *depends on the Parallel Postulate*.

It was not until the nineteenth century that an Italian mathematician named Eugenio Beltrami proved that Euclid's fifth postulate cannot be proven with the other four. This is not to say that the postulate is false, for Beltrami also proved that you cannot *disprove* the fifth postulate with the other four postulates. What he proved was that the Parallel Postulate was indeed *independent* of the other four. It is a marvel of mathematics that such a thing is even possible. How does one go about proving that a particular statement can be neither proven nor disproven? Beltrami accomplished this feat by providing an alternate model of geometry that satisfies all of the Euclidean postulates except the last.[85]

Think of it this way. Suppose we are trying to show that a fruit's being orange is independent of its being round and sweet. We could do this by demonstrating two things. First, we produce a piece of round, sweet fruit that *is* orange, for example a tangerine. Second, we produce a piece of round, sweet fruit that *is not* orange, for example an apple. Together, the tangerine and the apple show that a round, sweet piece of fruit can be either orange or not orange. In other words, being orange is *independent* from being round and being sweet. It also means that logically the properties of being round and sweet can never be used to prove that something is orange or not orange. Contrast that with trying to show that the ability to roll is independent from being round and sweet. Coming up with a round, sweet piece of fruit that can roll is not difficult—a tangerine will suffice—but coming up with a round, sweet piece of fruit that cannot roll proves to be impossible. That is because the ability to roll depends on at least one of our "postulates," in this case being round. In the case of geometry, being round and sweet means "satisfying the first four postulates," and being orange means satisfying the fifth postulate. The tangerine is the standard Euclidean model, and the apple is Beltrami's model.[86]

To see how a "different model" of geometry might be constructed, we return to the geometry on the surface of the sphere. At a basic level, points

85 This is a bit of an oversimplification. It was important that Beltrami defined his model in terms of a standard Euclidean model, but one for which the Parallel Postulate does not hold. Therefore, if Euclid's system is consistent (meaning that the five postulates do not imply any inherent contradiction), so is Beltrami's model. At that point we have two consistent models: one that accepts the Parallel Postulate and one that rejects it. Therefore, logically, it cannot be possible to prove the Parallel Postulate or its opposite with the other four postulates.

86 I am well aware of the limitations of this analogy. Even so, it will bear some fruit in helping to understand the concept of independence.

and lines in this spherical model are pretty much the same as their flat-plane counterparts. Points are exact locations, and lines are the things that join two points in as short a way as possible.[87] In the spherical model, though, some strange things happen, especially with triangles.

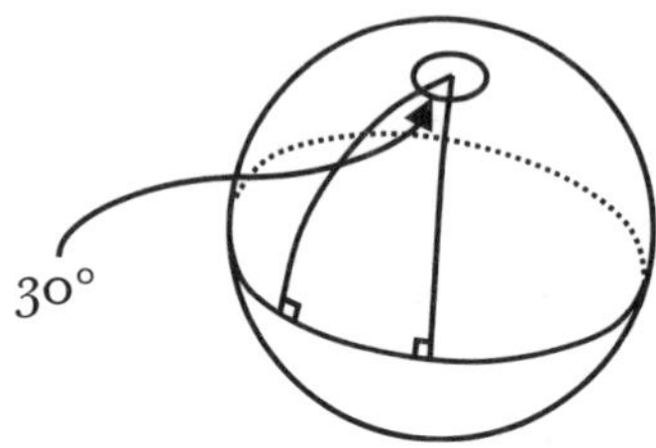

It may seem strange to think about this shape as a triangle, but remember that the objects in this geometry exist on the surface of the sphere. Thus, the sides of this triangle really are straight lines. Think about walking along the surface of the sphere. The path you would travel between two of the triangle's vertices is exactly the side of the triangle that joins them. It is the shortest path from the starting point to the ending point. To be clear, this path is *not* the same thing as a normal Euclidean line, *nor* is the triangle the same thing as a flat Euclidean triangle. That is precisely the point. This is a *different* geometry.

The particularly strange thing in this spherical geometry is the measure of the angles in the spherical triangle. The two base angles are formed by lines that start from the equator and head to the north pole. As such, they are both perpendicular to the bottom of the triangle. The top angle is labeled 30°, and so the sum of these angles is 210°! In spherical geometry, the angles of a triangle do not add to a straight line. In fact, they do not even add to the same thing for all triangles. We could spread the base of the triangle out at the equator, which would leave the two base angles as 90° but *increase* the top angle, making the sum now greater than 210°! The smaller the triangle becomes, the closer the sum of the angles is to 180°, but the sum will never actually be 180°. The precise reason this happens is that our new geometry does not follow the five postulates of Euclid. If it did, our proof that the angles of a triangle always add to 180° would

87 Officially, lines are *great circles*, or circles around the sphere that have the same radius as the sphere (or are "as big as possible"), and line segments joining two points are pieces of these great circles.

be valid. If these postulates do not hold in a given model, then all bets about the angle sum in triangles are off. We also now have the resolution to our opening riddle. A bear can travel south, east, and north and end up where he started if his departure point is the north pole. We leave it to you to work out the color of the bear.[88]

At this point, we should note that this model is not the one Beltrami used. Actually, our model violates more than just the Parallel Postulate. It may be helpful to review each of Euclid's five postulates to see how they match up with spherical geometry. It seems to satisfy the first—two points can be joined by a line—but some people have interpreted Euclid's first postulate as saying that two points can be joined by *one line and one line only*. If that is the case, then the postulate does not hold in our new geometry. Two points can be joined in *two* ways by going opposite directions around the sphere. The second postulate definitely holds: a line can be extended. That said, the extension ends up forming a circle that goes all the way around the sphere. The third postulate guaranteed that a circle can be drawn so long as we are given the center and the radius. This postulate does not hold in the case of spherical geometry. Of course, *some* circles can be drawn on the sphere; the problem is that not *all* circles can be drawn. Some will be too big for the sphere to contain them. This highlights an important distinction between our spherical geometry and Euclid's plane geometry. Euclid envisioned an infinite playing field, whereas the surface of our sphere is very much finite. The fourth postulate—that all right angles are congruent—*does* hold. But the pesky fifth postulate, the Parallel Postulate, does not hold. This, however, is difficult to see.[89]

88 There are actually an infinite number of solutions near the Antarctic. Start one mile north of a circle centered at the south pole with circumference $\frac{1}{n}$ for any whole number n. Traveling south one mile puts us on this circle. Traveling east one mile is equivalent to going around the circle exactly n times, putting us back where we were before our eastward journey. Traveling north one mile simply traverses the same path we took when going south, putting us back at our original starting point. Of course, the question of the color of the bear precludes us from being near the Antarctic since there are no bears there.

89 The Parallel Postulate can be replaced by another formulation named after the Scottish mathematician John Playfair. Playfair's Axiom says, "In a plane, given a line and point not on the line, there is at most one line through the point parallel to the given line." It also can be replaced with a stronger version: "In a plane, given a line and point not on the line, there is *exactly one* line through the point parallel to the given line." This formulation is so famous that it is often mistakenly labeled as the Parallel Postulate. I debated whether to include Playfair's Axiom in this chapter, but the difficulty of describing its equivalence to Euclid's original postulate distracted from the main purpose of the chapter. Admittedly, using this formulation, it is much easier to see that spherical geometry violates the postulate/axiom. On a sphere, there are *no* parallel lines. When extended, all lines are great circles, so they all meet at some point—two points, actually. Therefore, in spherical geometry, given a line and point not on the line, there are no parallel lines passing through the point that are parallel to the given line. Even though this formulation is easier to see, it was not worth the tangent of the two formulations of the Parallel Postulate. Thus, I regretfully relegate it to a footnote.

Thus, our spherical model demonstrates the consequences of systems that do not obey the Euclidean postulates, but unfortunately it does not prove the independence of the Parallel Postulate. We needed a round, sweet piece of fruit that is not orange. What we produced here is something like a sour piece of green fruit. To really prove the independence of the Parallel Postulate, we need a model that *obeys* the first four postulates but *violates* the fifth. Beltrami produced that model.

An oversimplified version of Beltrami's model is one that exists on a curved surface that looks like a horse saddle, but an infinite one.[90]

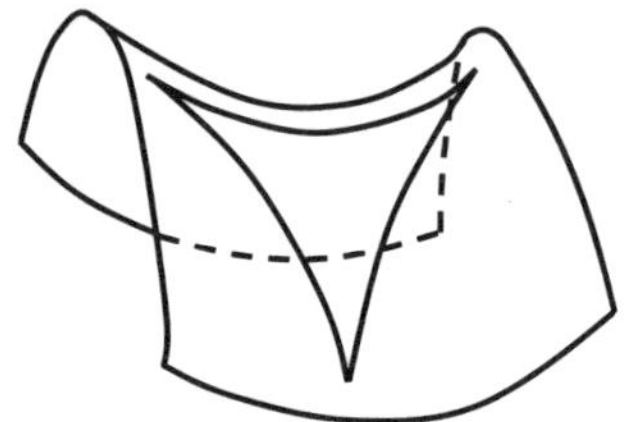

While Beltrami constructed a brilliant model based on Euclid's first four postulates, the idea of this sort of geometry was not new. The term for a geometry that obeys the first four postulates but adopts the opposite of the Parallel Postulate is *hyperbolic geometry*, and sometimes even *Lobachevskian geometry*, named after the Russian mathematician Nikolai Lobachevski, who studied these things. The more general term "non-Euclidean geometry" is also used, but that description could also apply to our spherical geometry. Even though others had worked on these non-Euclidean ideas, Beltrami constructed a particular model of a hyperbolic geometry that was consistent, or not self-contradictory. Think of it this way. It is one thing to posit the existence of a round, sweet piece of fruit that is not orange, but it is another thing to produce an actual apple. Beltrami produced the apple.

In the context of our mathematical journey thus far, there are at least two things about hyperbolic geometry that are worth noticing. First, while in spherical geometry the angles in a triangle add to more than 180°, in hyperbolic geometry the angles of a triangle always add to *less* than 180°. Second, in the Euclidean plane, there are only three regular

90 It is actually the case that the "saddle" appears when you zoom in on hyperbolic space. Beltrami's model uses the interior of a unit circle as the entire hyperbolic plane with chords as the hyperbolic "lines." This level of sophistication is well beyond the scope of this text.

polygons that can fit together nicely in a tile pattern: the equilateral triangle, the square, and the regular hexagon. In hyperbolic geometry, *any* regular polygon can be made to tile the hyperbolic plane.[91]

With Beltrami's creation, together with the regular Euclidean plane, we have two different consistent models of geometry that satisfy Euclid's first four postulates. They are both round and sweet pieces of fruit. The difference is that Euclid's plane satisfies the Parallel Postulate (it is a tangerine), and Beltrami's model satisfies the opposite of the Parallel Postulate (it is an apple). Because they are both legitimate models, it must be that the Parallel Postulate (the property of being orange) is independent of the other four postulates. Said differently, if the Parallel Postulate could be proven from the other four, Beltrami's model could not exist.[92]

Here at the end of this chapter, it is worth pausing to revel in the glory of what has been accomplished. We have seen throughout our journey the remarkable ability mathematics has for proving things. We now know that mathematics is even capable of proving that some things are neither provable nor disprovable.

The Independence of the Parallel Postulate
Euclid's first four postulates cannot be used to prove that the fifth postulate is true, nor can they be used to prove that the fifth postulate is false.

91 The polygons that "tile the sphere" are those that are related to the Platonic solids. For example, the pentagons in the icosahedron can be "projected out" to a surrounding sphere to produce a tiling pattern on that sphere. The same can be done with the other solids. On a sphere, however, we can also get some strange tilings: two n-gons (for example, two 13-gons) that meet at the equator, or even n "2-gons." The concept of a "2-gon" does not make sense in the Euclidean plane, but it does on the sphere. Drawing, for example, ten equally spaced lines of longitude on the Earth will form twenty "regular 2-gons."

92 To be fair, we also need an actual model for the Euclidean set of five postulates. I have waved my hands here and said "the Euclidean plane" is such a model. Some work needs to be done to build out this model, but flat-plane geometry is so intuitive to most readers that such a construction is not necessary in this context.

CHAPTER 4
IMPOSSIBLE CONSTRUCTIONS

Euclid offers two types of proofs in the *Elements*: the "showing" kind and the "doing" kind. In the showing kind of proof, Euclid claims something is true about a geometrical object and then proceeds to offer a logical explanation for why it is true. The proofs of the Triangle Angle Sum Theorem and the Pythagorean Theorem are of this sort. At the end of such proofs, Euclid writes, "Q.E.D." This Latin acronym stands for *quod erat demonstrandum*—literally "that which was to have been demonstrated," but usually taken to mean "that which was to have been demonstrated has now been demonstrated."

The "doing" type of proofs are quite different, and they bear the mark "Q.E.F." at the end, standing for *quod erat faciendum*—"that which was to have been done" or "that which was to have been done has now been done." The propositions of this type are *tasks to be accomplished* or even *things to be built*. For this reason, they are called *constructions*. In fact, the very first proposition in the *Elements* is of this sort: "On a given finite straight line to construct an equilateral triangle." The task is to draw an equilateral triangle given one side length. Fair enough, but what are the "rules" of drawing, and what are the tools? Commonly, Euclidean constructions are called "compass and straightedge constructions" because they are to be accomplished using only those two tools. But this does not tell the entire story behind these tasks. It is not so much the tools that are important; it is what the tools themselves make.

We saw in the previous chapter that the entirety of the *Elements* is built off of five postulates. Recall that the first three are really "doing" postulates. They claim that three things can be done. In modern parlance, these are the following:

1. Two points can be connected with a straight line.
2. A straight line can be extended in either direction indefinitely.
3. Given a center and a radius, a circle can be drawn.

For the Greeks, there were two basic geometrical objects in the universe: lines and circles. The first two postulates say that both *finite* lines and *infinite* lines can be drawn.[93] The third postulate says that circles can be drawn. The combination of these three is precisely why the compass and the straightedge became the geometer's tools. The compass is the "circle-drawing tool," and the straightedge is the "line-drawing tool."

There were at least two reasons why Euclid was concerned with constructions. One is that we often need to add certain elements to a diagram to move a particular proof along. The proof of the Triangle Angle Sum Theorem needed a parallel line going through a point, and the Pythagorean Theorem needed several auxiliary lines to be added to the diagram. To ensure that his proofs were complete, Euclid went to great lengths to show that these sorts of additions actually could be accomplished. The second reason, however, is perhaps even more important. If the circle and line are the two basic elements in the geometrical universe, we should be able to use them to construct other geometrical objects. To offer an analogy, for the Greeks the circle and the line are like the "prime numbers" of geometry. It is significant that the very first proposition Euclid offers is to use these "prime shapes" to construct another shape—the equilateral triangle—and that the *Elements* ends with the construction of the Platonic solids.

How *does* Euclid use circles to draw an equilateral triangle? He begins with a finite straight line.[94]

A ——— *B*

He then draws a circle using *A* as a center and the line as the radius.

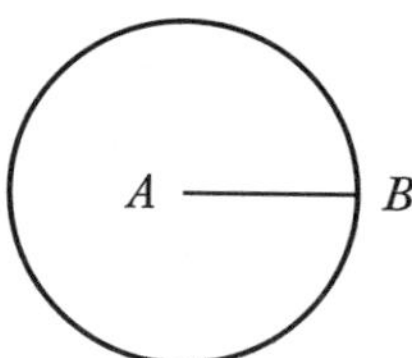

Euclid then does something brilliant: he draws a second circle using *B* as the center but still using the same radius.

93 There is some debate about the interpretation of the first two postulates, which has something to do with the potential infinite versus the actual infinite. Some people take the second postulate to mean only that a line can be extended as far as we need, which would make the extension in any particular case still finite. Others take it to mean that actual infinite lines exist.

94 It is a good reminder that for Euclid, as for us in this text, there is not a distinction between a line (which in modern terminology is infinite) and a line segment (which in modern terminology is finite).

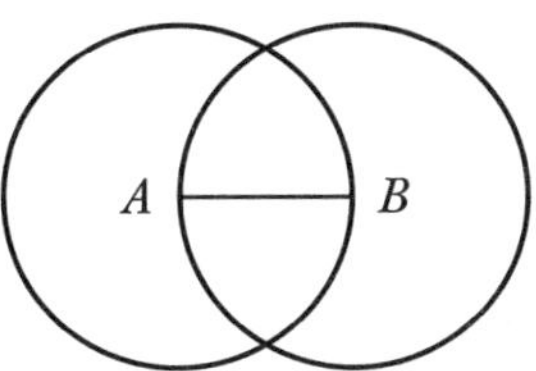

Finally, he uses the intersection of the circles as the third point for a triangle.

Not content with just drawing the shape, Euclid then goes on to prove that the construction does what it claims to do. The proof hinges on the three sides of the triangle each being a radius of at least one of the two circles and all radii of a circle being equal.[95] Because the two circles have the same radius—after all, we drew them that way—all three sides must be equal. *Q.E.F.* It is beautiful, it is simple, and it is beautiful because of its simplicity.

Throughout the *Elements*, Euclid constructs shape after shape. He constructs parallel lines, perpendicular lines, squares, and even three-dimensional shapes.[96] He divides lines into various proportions, splits angles exactly in half, and copies angles from one location to another. In all of his work, however, three constructions notably eluded Euclid. In fact, they eluded all mathematicians for thousands of years.

1. *Trisecting the Angle:* Given any angle, divide it equally into three.
2. *Squaring the Circle:* Given a circle, draw a square that has the same area.
3. *Doubling the Cube:* Given a cube, draw another cube that has twice the volume.

95 Euclid uses the equality of radii as the very definition of a circle: "A circle is a plane figure contained by one line such that all the straight lines falling upon it from one point among those lying within the figure are equal to one another."

96 It is here that we realize that "paper" is not a tool of the geometer in the same way that the compass and straightedge are. Of course, a three-dimensional object could never be drawn with a compass and a straightedge on flat paper, but the ability to "make" the sides, even in space, is still important for Euclid.

The last problem has an interesting legend behind it. The ancient town of Delos was plagued with rampant civil unrest.[97] Seeking a solution to the political violence, the citizens consulted the oracle of Apollo, who told them to build a new altar to the god that was twice the size of the existing altar located prominently in the center of the town. The altar was a perfect cube, and therefore the task was to double this cube. Unfortunately for the citizens of Delos, the difficult task of damping down the tumult of the people was replaced by an equally challenging geometry problem. Apparently being big on consultations, they went next to the philosopher Plato. In true form, rather than showing them how to perform the construction, he instead reinterpreted the oracle's solution and told the citizens that they would be better off studying geometry than fighting amongst themselves. Because of this legend, the problem of doubling the cube has been known as the Delian problem.

It is actually surprising that each of these three tasks proved to be so difficult. After all, in Proposition 9 of Book I, Euclid shows how to *bisect* any angle, meaning to divide it into two equal parts, and the construction is rather tame. In fact, using his construction, we could cut an angle into *four* equal parts by first bisecting the angle and then bisecting each part. If *two* parts is easy and *four* parts is easy, why is it so difficult to divide an angle into *three* equal parts?[98]

Moreover, in Proposition 14 of Book II Euclid shows how to take any polygon and construct a square that has the same area. This is incredible! We can take any polygon we want, even one with 1,789,211 sides all of different lengths, and we can use circles and lines to construct a square that has the exact same area as the 1,789,211-gon. If we can accomplish the monumental feat of squaring any polygon, why is it so difficult to square the circle?

Finally, it is easy enough to double a line (to take a given line and construct a second one with twice the length) and even easy enough to double a square (to take a given square and construct a second one with twice the area). Euclid does not specifically do the latter, but the construction

97 In some versions of the story, Delos was plagued by an actual plague.

98 Note that we can trisect *some* angles. In fact, we can easily trisect a right angle. Given that we can construct an equilateral triangle, this means that a 60° angle is constructible. Using Euclid's method, we can bisect the 60° angle to produce a 30° angle. This 30° angle can then be copied onto the right angle, which effectively trisects it. The hard task is to come up with a construction that works for any angle, regardless of its measure or even of *knowing* its measure.

follows in any number of ways from some of his other accomplishments. A particularly elegant way to double the square it is to construct a square on the diagonal of the original square. (Note that Euclid shows that a square, like an equilateral triangle, is constructible on any given line.)[99] Once the larger square is constructed along the diagonal, we can add a few dotted lines to see that the big square is twice the area of the small square. This argument resembles the one that Plato presents in the *Meno*, when Socrates teaches the slave boy about this very task of doubling a square.[100]

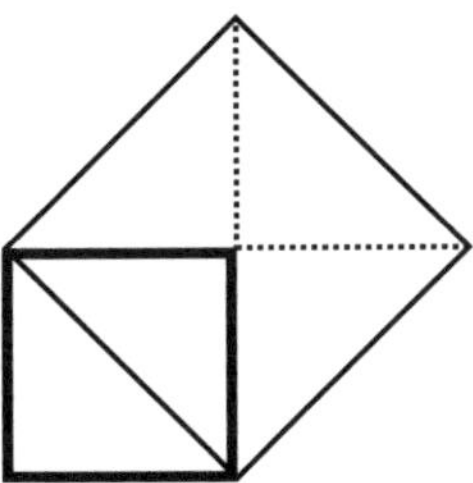

If a line and a square can be doubled so easily, why is it so difficult to double a cube?

The German mathematician Johann Carl Friedrich Gauss made a major advance in compass and straightedge constructions. His first contribution deals with the construction of regular polygons. I noted above that Euclid was able to construct two of these regular polygons: the equilateral triangle and the square. In Book IV, he gives an exquisite construction of a regular pentagon and culminates the book in the construction of a regular pentadecagon (fifteen sides). In 1796, when Gauss was only nineteen years old, he proved that the regular heptadecagon (seventeen sides) was constructible.[101]

Gauss's greatest contribution to constructions goes well beyond the regular heptadecagon. Five years later, Gauss proved an entire collection

99 Euclid shows that squares are constructible in Proposition 46 of Book I, just before the Pythagorean Theorem. Remember, the Pythagorean Theorem is a claim about squares that are drawn on the sides of a right triangle, so it is important for Euclid to show that such a task is even possible. Of course, it is not merely for use in the Pythagorean Theorem. The construction of a square is also important as an example, together with the equilateral triangle from Proposition 1, of a constructible regular polygon.

100 What Plato actually does (through Socrates) is to surround the small square with the larger square such that the vertices of the small square are the midpoints of the large square's edges, but the proof still relies on counting the small triangles that appear in each square. I prefer the above version because of the ease of constructing the large square on the diagonal of the small square.

101 Interestingly enough, Gauss only proved that a regular heptadecagon could be constructed, but he did not actually provide a construction. That task was accomplished by a few mathematicians in the years that followed Gauss's proof. Johannes Erchinger and Herbert William Richmond are credited with developing valid constructions.

of regular polygons to be constructible. His work relies on numbers as much as shapes, using a type of prime number known as a *Fermat prime*.

DEFINITION: A Fermat prime is a prime number that is in the form $2^{2^n}+1$.

The first Fermat prime is 3 because $3=2^{2^0}+1$, and the second is 5 because $5=2^{2^1}+1$. The next several Fermat primes are in the following table.

n	$2^{2^n}+1$
0	$2^{2^0}+1=3$
1	$2^{2^1}+1=5$
2	$2^{2^2}+1=17$
3	$2^{2^3}+1=257$
4	$2^{2^4}+1=65{,}537$

With the first five examples being prime numbers, we are tempted to think that $2^{2^n}+1$ is always prime. Fermat originally thought this to be the case, but Euler later showed that substituting 5 in for n will result in a number that is not prime.[102] This raises the question, "How many Fermat primes are there?" Amazingly enough, we do not know. In fact, to date, we are only aware of *five* of them: the ones listed in the table.

Gauss showed that you can multiply any number of Fermat primes together (so long as you don't repeat) and then multiply by 2 as many times as you wish (including no times at all if you so desire), and the result will represent a constructible regular polygon.[103]

Gauss's Theorem About Constructible Regular Polygons A regular n-gon is constructible if n is the product of any number of 2s as well as distinct Fermat primes.

102 Lest we think that it was an easy task to show that the case of $n=5$ did not produce a prime, $2^{2^5}+1$ is 4,294,967,297, which is $641\times 6{,}700{,}417$.

103 This can be described in terms of prime factorization. A regular n-gon is constructible if the prime factorization of n uses only 2s and Fermat primes, with the Fermat primes appearing at most once.

For example, the heptadecagon is constructible because it is a Fermat prime. (We are allowed to use only one Fermat prime, and we don't have to use any factors of 2.) The regular 34-gon is also constructible because $34=17\times 2$, as is the regular 68-gon because $68=17\times 2\times 2$. Moreover, because 17 and 5 are both Fermat primes, the regular 85-gon is constructible because $85=17\times 5$, and so is the regular 170-gon because $170=18\times 5\times 2$. But we don't *have* to use Fermat primes. We can use only 2s. Therefore, the regular octagon is constructible because $8=2\times 2\times 2$.

It is worth noting that the Greeks knew about part of this theorem. They knew that regular polygons with three sides (the first Fermat prime), five sides (the second Fermat prime), and even fifteen sides (the product of the first two Fermat primes) were all constructible. We already noted that Euclid himself provides constructions for all three of these cases. They also knew that if a regular polygon was constructible, so was one with twice the number of sides, which takes care of the "any number of 2s" part of Gauss's theorem.

Gauss went one step further, however, and stated that these are the *only* constructible regular polygons. For example, the regular heptagon (seven sides) is *not* constructible because it is not the product of 2s and Fermat primes, and neither is the regular nonagon (nine sides). Gauss conjectured this but never published a proof. A little over three decades later, the French mathematician Pierre Wantzel proved this second part of Gauss's theorem. The combined theorem bears both mathematicians' monikers.

Gauss–Wantzel Theorem
If n is the product of any number of 2s together with distinct Fermat primes, then a regular n-gon is constructible. Moreover, these are the only constructible regular polygons.

How is all of this related to the three classically "difficult" constructions? First, and not incidentally, mathematics now had a demonstration that certain constructions cannot be accomplished. This gave hope that it might be possible to prove that the three classical tasks are not just difficult but actually impossible. In point of fact, it did not take long. In the paper in which Wantzel proved that the non-Gaussian regular polygons were impossible to construct, he used the same techniques to show that it is impossible to trisect an angle and to double the cube. The proof that it was impossible to square the circle came a bit later but used the same ideas and was published by another German mathematician, Carl Louis Ferdinand von Lindemann. Thus, the task so valiantly attempted by the Greeks and other mathematicians throughout centuries proved to be impossible.

> *Impossible Constructions Theorem*
> It is not possible to trisect a given angle, square a given circle, or double a given cube.

How do these proofs work? How is it possible to prove that something is impossible? It was a monumental moment for mathematics that required knowledge of some very modern concepts and techniques, and yet we can get a glimpse of the proof by sketching the main ideas.

The real moment of genius was to focus on *numbers* instead of *shapes*. Focusing on numbers is the precise reason why the constructability of regular polygons and all three classical construction problems were conquered at nearly the same time. The whole subject of constructible objects was reframed into a question about line lengths.

Given a line of length 1, what other possible lengths can be constructed?

In the chapter on the existence of irrational numbers, I mentioned that Euclid knew how to add two lines side by side. This means that starting with a line with length 1, we can construct line with any whole number length. For example, we can draw a line of length 4 because $4=1+1+1+1$. I also noted that Euclid knew how to "multiply" segments together and even "divide" them. In other words, if we can construct two segments of lengths a and b, then we can also construct segments of lengths $a \times b$ and $a \div b$. Do you see what is going on here? We are developing a sort of "algebra of constructions." The construction for division was particularly important because this means that segments of any rational length were constructible. But we also know that we can construct some segments with irrational lengths, for example $\sqrt{2}$. This "algebra of constructions" allows us to define an entirely new category of numbers: the *constructible numbers*.

> DEFINITION: A number is a *constructible number* if a segment of that length can be drawn using only a straightedge and a compass.

Gauss's work is brilliant because it allowed us to figure out precisely which numbers are constructible and which numbers are not. It turns out that addition, subtraction, multiplication, division, and "square rooting" are the *only* ways to make more constructible numbers.[104]

> THEOREM: Constructible numbers are precisely those that can be formed by combining whole numbers with addition, subtraction, multiplication, division, and square roots.

These few operations advance us fairly far down the road, by the way. For example, the complicated number $\frac{\sqrt{3-\sqrt{2}}}{4}$ is constructible because 2, 3, an 4 are constructible, so we can combine square roots, subtraction, and division to get the desired result. The surprising thing, though, is that *not all real numbers are constructible.* Importantly, Wantzel showed that the cube root of 2, i.e., $\sqrt[3]{2}$, was not constructible because we can never create it starting with the integers, the four basic operations, and square roots. In fact, he showed that none of the cube roots are constructible (except "nice" ones such as $\sqrt[3]{8}$, which equals 2). In other words, we have found a category of numbers that contains not only the rational numbers but also some irrational numbers.

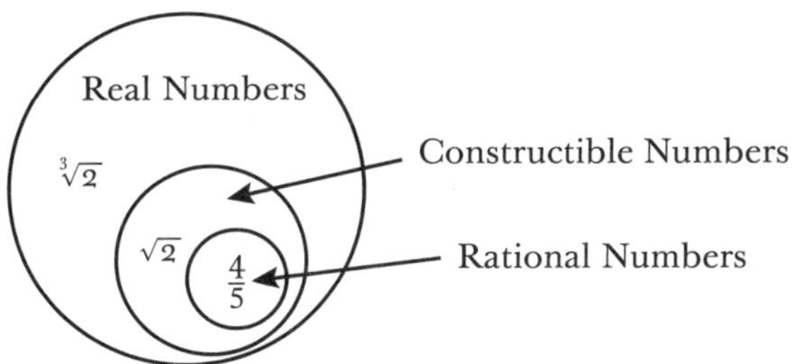

All rational numbers are constructible, some irrational numbers are constructible ($\sqrt{2}$), and some irrational numbers are not constructible ($\sqrt[3]{2}$).

104 For mathematical reasons, constructible numbers actually include the negatives of their positive counterparts. For our purposes, that is not relevant, and we are content with the simpler definition using only positive numbers.

Wantzel's identification of $\sqrt[3]{2}$ as a non-constructible number was not arbitrary. This number is important for doubling the cube. If doubling the cube were possible, then we could certainly double the cube with side length 1. A cube with side length 1 has a volume of 1, so a doubled cube would have a volume of 2, which would give it a side length of $\sqrt[3]{2}$.

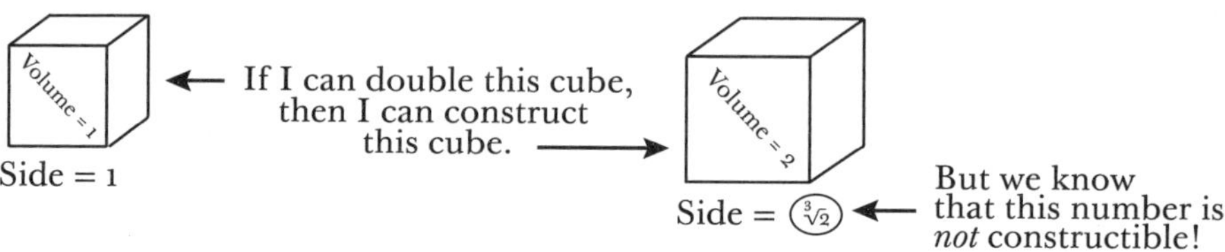

In short, if doubling the cube were possible, then we could construct a segment with length $\sqrt[3]{2}$, meaning $\sqrt[3]{2}$ would be constructible. Knowing that $\sqrt[3]{2}$ is *not* constructible means that doubling the cube is *not* possible. (This is another proof by contradiction.) The proofs that the other two constructions are impossible work exactly the same way: *if* the construction is possible, *then* this-particular-number is constructible, but we know that this-particular-number is *not* constructible. Therefore, the construction is not possible. The particular number—the fly in the ointment—for doubling the cube is $\sqrt[3]{2}$. For squaring the circle, the particular number is π. For trisecting the angle, it is the number cos(20°).[105] The difficult work here is in showing that these three numbers ($\sqrt[3]{2}$, π, and cos(20°)) cannot be the result of the combining whole numbers, the four basic operations, and the square root.[106]

The first amazing thing is that these problems, long known to be difficult, turn out to be more than difficult. They are, in fact, impossible. The second amazing thing is that their impossibility was shown by thinking about geometrical constructions as numbers. It is a rare and historically profound moment when two areas of mathematics are joined, in this case geometry and number theory. Who would have thought that trying to draw regular polygons has anything to do with prime numbers? Well, Gauss did.

105 As a reminder from trigonometry, cos(20°) is the length of the longer leg in a right triangle with a 20° angle and a hypotenuse of length 1.

106 The proof of the impossibility of constructing the regular polygons relies on the same technique: identifying a non-constructible number that would appear if such a construction were possible.

CHAPTER 5
A FORMULA FOR THE FIBONACCI NUMBERS

THE FIBONACCI numbers are so famous that it is hard to know how best to introduce them. We could honor their namesake, the Italian mathematician, and introduce them using the growth of rabbit populations.[107] Or we could honor their appearance nearly two millennia earlier in Indian poetry and introduce them as they relate to the number of ways to combine long and short syllables.[108] Given the prominent role that Euclid has played in our journey thus far, we could introduce them as they related to his algorithm for finding the greatest common divisor of two numbers.[109] Or we could introduce the spiral patterns in plants and other natural phenomena that exhibit the Fibonacci numbers. With such numerous and varied contexts and the fact that the numbers are so well known, we opt instead simply to write them down, in all their beauty, with neither pomp nor circumstance.

$$1, 1, 2, 3, 5, 8, 13, 21, 34, 55, 89, \dots$$

The sequence begins with two 1s, and every number after that is found by adding the previous two. The idea is so simple, and yet both mathematicians and mother nature herself seem to have a great love for these simple numbers. The formula "the next number is equal to the sum of the previous two" is known as a *recursive formula* because it depends on knowing earlier values to calculate later values. There is nothing mathematically wrong with this sort of formula, and in fact the Fibonacci formula is not alone among recursive formulas in its elegance. But it is inconvenient if we want, say, the 500th Fibonacci number. In order to calculate this, we would need both the 499th and 498th Fibonacci numbers, and to

107 We start with one pair of rabbits. Each pair of rabbits will mate at the age of one month and then produce exactly one other pair of rabbits at the end of every other month after that. If the rabbits never die, Fibonacci asks, "At the end of n months, how many pairs of rabbits will there be?" The population counts at the end of each month are the Fibonacci numbers.

108 If a long syllable takes two units of time and a short syllable takes one unit of time, how many possibilities are there for a line length of six units? For example, we could have three long units; or we could have one long, two short, and one long; or we could have one short, two long, and one short. There are thirteen possibilities, which is a Fibonacci number. For other line lengths, we get the rest of the Fibonacci numbers.

109 Euclid's algorithm for finding the greatest common divisor of a and b is to use long division to write b (the greater of the two) as $b = a \times q + r$, then rinse and repeat with a and r. If ever we get a remainder of 0, the penultimate remainder is the greatest common divisor. The algorithm is *fast*, meaning you can start with fairly large numbers and find the greatest common divisor in a relatively small number of steps. The slowest scenario (which is still pretty fast) for Euclid's algorithm is when the original numbers are consecutive Fibonacci numbers.

calculate those we would need the 497th and 496th Fibonacci numbers, and so on. In other words, we would have to build them up one at a time starting with 1 and 1 until we finally arrived at the 500th number.[110]

Some recursive sequences have a straightforward *explicit formula* that allows for more efficient calculation of larger values. For example, consider a sequence formed by simple addition.

$$2, 5, 8, 11, 14, 17, 20, 23, 26, 29, 32, \ldots$$

Each term is three greater than the one that comes before it. A recursive formula for this sequence would be "the next number is equal to the previous number plus 3." Stated that way, we would still have a task ahead of us in determining the 500th number. This particular sequence of numbers, however, has an explicit (non-recursive) formula. The nth number is always $2+3\times(n-1)$. This formula makes sense: we start with 2 and then "add a bunch of 3s" to get the subsequent numbers. Having an explicit formula allows us to calculate the 500th without having to calculate the previous 499 numbers.

$$2 + 3 \times (500-1) = 1{,}499$$

This simpler sequence depends on only one previous term, whereas the Fibonacci sequence depends on *two* previous terms. And yet, how much more complicated could that possibly make an explicit formula for the Fibonacci sequence? It turns out that it is quite a bit more complicated. The formula is known as Binet's Formula, named after the French mathematician Jacques Philippe Marie Binet.[111]

Binet's Formula for the Fibonacci Numbers

The nth Fibonacci number can be calculated with this formula:

$$\frac{\left(\frac{1+\sqrt{5}}{2}\right)^n - \left(\frac{1-\sqrt{5}}{2}\right)^n}{\sqrt{5}}.$$

110 The 500th Fibonacci number is 139,423,224,561,697,880,139,724,382,870,407,283,950,070,256,587,697,307,264,108,962, 948,325,571,622,863,290,691,557,658,876,222,521,294,125.

111 While the formula is attributed to Binet, it seems that it was already known to Abraham de Moivre and Daniel Bernoulli about a hundred years earlier.

The proof that the formula is correct is difficult and beyond what we can cover in this book, but the beauty of this formula is incredible and is worth exploring for a variety of reasons. To begin with, it is amazing that Binet's formula produces whole numbers at all. Look at the details of the formula. The number $\sqrt{5}$ shows up three times in a complicated fraction of fractions, twice with an exponent. Keep in mind that like $\sqrt{2}$, the number $\sqrt{5}$ is irrational. We might reasonably assume that Binet's expression would produce a slew of irrational numbers, and that even if it managed to escape the irrational numbers it would at least produce non-integer fractions. And yet, so long as n is a positive integer, the formula always gives us a whole number, and *specifically gives us the whole numbers* 1, 1, 2, 3, 5, 8, 13, 21, 34, 55, 89, ...

The presence of an irrational number, $\sqrt{5}$, is immediately apparent and interesting. In our other, simpler sequence, the two numbers appearing in the explicit formula were 2 and 3, and both numbers had clear ties back to the sequence as the *first term* and *common difference* between terms. What in the world does $\sqrt{5}$ have to do with "start with 1 and 1, then add the two previous terms to get the next term?" As interesting as the presence of $\sqrt{5}$ is in Binet's formula, the trained mathematical eye would be drawn to another number: $\frac{1+\sqrt{5}}{2}$. This is a famous number known as the *golden ratio*, sometimes even called the "divine proportion" (*proportio divina*). The number is often denoted φ and is approximately equal to 1.618, though like π and $\sqrt{2}$ it is irrational. As such, its decimal expansion neither repeats nor terminates. It was called the divine proportion in part because of its aesthetically pleasing nature: a rectangle whose sides are in the golden ratio is said to be the most beautiful of all rectangles. When constructing such rectangles, the golden ratio shows up in a recursive way. If we cut off a square from a golden rectangle, another golden rectangle is left over.[112]

112 This relationship in the golden rectangle can help us prove that the golden ratio is $\frac{1+\sqrt{5}}{2}$. Notice in the picture that follows, $b = a + c$ because the left part of the rectangle is given as a square. Therefore, because $\frac{b}{a} = \frac{a}{c}$ it must be that $\frac{a+c}{a} = \frac{a}{c}$. The left side fraction can be split: $\frac{a}{a}+\frac{c}{a}=\frac{a}{c}$, or $1+\frac{c}{a}=\frac{a}{c}$. Because $\frac{a}{c}$ is the golden ratio, or φ, the fraction $\frac{c}{a}$ is the reciprocal of the golden ratio, or $\frac{1}{\varphi}$. Thus, $1+\frac{c}{a}=\frac{a}{c}$ is the same as $1+\frac{1}{\varphi}=\varphi$. Multiply both sides by φ to get $\varphi+1=\varphi^2$. Solving this quadratic equation gives the solutions $\frac{1\pm\sqrt{5}}{2}$. But the golden ratio is a positive number, so $\varphi=\frac{1+\sqrt{5}}{2}$.

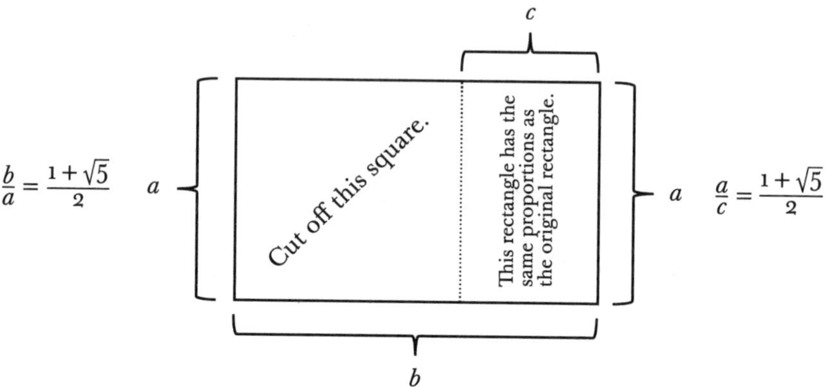

This golden rectangle is so well known, or at least well intuited, that it shows up in art and architecture frequently. Euclid himself knew of this number and in fact demonstrates how to construct it in Book II of the *Elements* by cutting a line into two segments that are in the golden ratio. His demonstration means that the golden ratio is constructible[113] and that the golden rectangle is constructible.

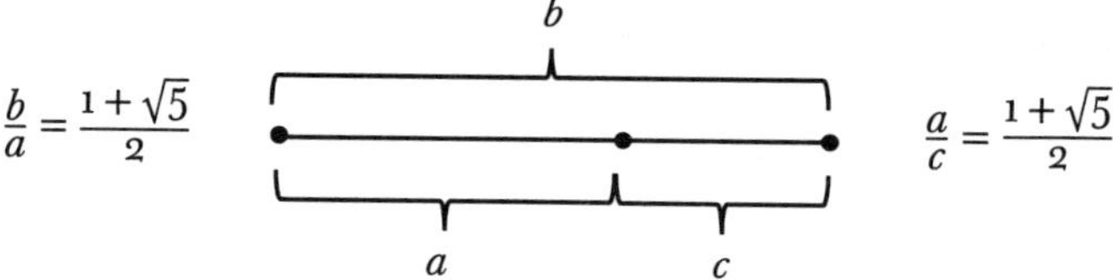

It seems, however, that the most important application of the golden ratio for Euclid occurs in the construction of the regular pentagon, which I have alluded to a few times already. We will not go through Euclid's actual construction, but it is worth noting how prevalent the golden ratio is in the regular pentagon. Each of the following pairs of lengths is in the same ratio as $\varphi = \frac{1+\sqrt{5}}{2}$: *DE* and *EX*, *EX* and *XY*, *UV* and *XY*, *EY* and *EX*, and *BE* and *AE*.

113 From our work in the previous chapter, we already know that is constructible because can be built from integers, the four basic operations, and square roots.

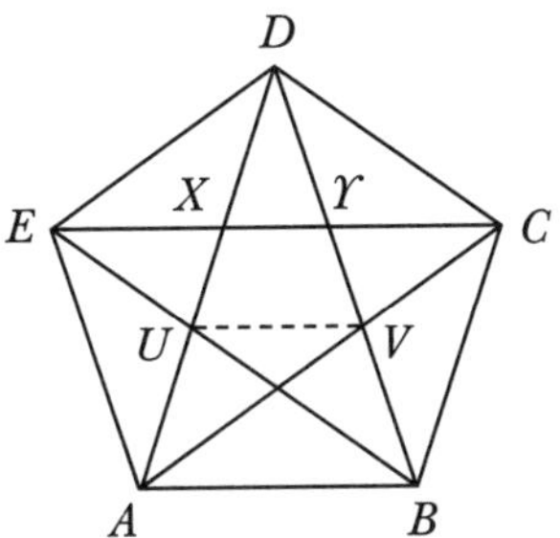

In addition to its geometrical beauty, the numerical beauty of φ can be seen in one of its continued fraction representations. We saw earlier that π, whose digits seem to be without pattern, is equal to several continued fractions that clearly have patterns.

$$\pi = 3 + \cfrac{1^2}{6 + \cfrac{3^2}{6 + \cfrac{5^2}{6 + \cfrac{7^2}{6 + \cfrac{9^2}{6 + \cfrac{11^2}{6 + \cdots}}}}}}$$

The golden ratio has an even more elegant continued fraction.[114]

$$\varphi = \frac{1+\sqrt{5}}{2} = 1 + \cfrac{1}{1 + \cfrac{1}{1 + \cfrac{1}{1 + \cfrac{1}{1 + \cfrac{1}{1 + \cfrac{1}{1 + \cdots}}}}}}$$

114 One painful omission from this discussion, which I relegate here to a footnote using admittedly too few details, is that the golden ratio is "the most irrational of all numbers." The reason for this has precisely to do with its continued fraction representation. If you pause to think about these continued fractions, they are attempts to write an irrational number as its integer part plus a small leftover, i.e., "1 over something," and then to do the same with that "something." *Simple* continued fractions (with all numerators of 1) are the *fastest* way to approximate irrational numbers. In other words, they produce the rational approximations that are *closest* to the irrational number. The "slowest" of these "fastest" approximations would occur when the "leftover" is as big as possible, and that happens precisely when the diagonal numbers are also 1. When *all* of the diagonal numbers are 1, we end up with the golden ratio. The consequence of this is that while all irrational numbers "avoid" being rational, the golden ratio avoids it more than any other number. It is the "most irrational" of all of the irrational numbers.

All of this said, what is important for us is the connection that φ has with the Fibonacci numbers. With that, let's return to Binet's formula.

$$\frac{\left(\frac{1+\sqrt{5}}{2}\right)^n-\left(\frac{1-\sqrt{5}}{2}\right)^n}{\sqrt{5}}$$

We have already observed that the first number in the numerator is the golden ratio, φ, but the second number, $\frac{1-\sqrt{5}}{2}$, can be written in two different ways using φ: $-\frac{1}{\varphi}$ and $1-\varphi$. (The algebra for this is not terribly complicated, but I will leave it for you to work out.) Using these two representations, we can write Binet's formula in two ways involving the golden ratio.

$$\frac{\varphi^n-\left(-\frac{1}{\varphi}\right)^n}{\sqrt{5}} \text{ or } \frac{\varphi^n-(1-\varphi)^n}{\sqrt{5}}$$

The presence of φ in Binet's formula leads to a well-known property of the Fibonacci numbers: their consecutive ratios will become closer and closer to the golden ratio.[115]

Fibonacci Number	Next Fibonacci Number	Ratio	Decimal
1	1	$\frac{1}{1}$	1
1	2	$\frac{2}{1}$	2
2	3	$\frac{3}{2}$	1.5
3	5	$\frac{5}{3}$	1.66666666...
5	8	$\frac{8}{5}$	1.6
8	13	$\frac{13}{8}$	1.625
...	...	...	...
46,368	75,025	$\frac{75{,}025}{46{,}368}$	1.618033...

115 These approximations, while seemingly close, do not converge as quickly as we may like. This slow convergence is related to the previous footnote about the golden ratio being the "furthest away" from being rational, and both ideas are related to the worst-case scenario for Euclid's algorithm for finding the greatest common divisor of two numbers.

A full justification of this property is another adventure for another time, but one way to accomplish it is to use Binet's formula.[116] Like many mathematical achievements, his remarkable and exquisite formula builds a bridge between two seemingly unrelated ideas. In this case, it establishes a deep connection between one of the most important sequences in all of mathematics, the Fibonacci sequence, and one of the most important numbers in all of mathematics, the golden ratio.[117] Even without a full proof, the formula is worthy of a privileged place in mathematical lore.

116 Another way to show this is to start with the ratio of two consecutive Fibonacci numbers: $\frac{F_{n+1}}{F_n}$. As n becomes larger, this ratio will get closer to some value, which we can call x. (We actually should prove that fact, but that requires some calculus.) We can then rewrite the expression using the Fibonacci relationship and split apart the resulting fraction: $\frac{F_{n+1}}{F_n} = \frac{F_n+F_{n-1}}{F_n} = \frac{F_n}{F_n} + \frac{F_{n-1}}{F_n} = 1 + \frac{F_{n-1}}{F_n}$. The expression $\frac{F_{n-1}}{F_n}$ is the ratio of consecutive Fibonacci numbers, except that it is upside down from our original relationship. No worries: this just means that the fraction $\frac{F_{n-1}}{F_n}$ approaches $\frac{1}{x}$ as n becomes larger. Therefore, the expression $1+\frac{F_{n-1}}{F_n}$ approaches $1+\frac{1}{x}$. We now have that $x = 1+\frac{1}{x}$. Solving this equation will demonstrate that x is the golden ratio.

117 The golden ratio is one specific type of *metallic ratio*. The golden ratio describes certain rectangles that, when a square is removed, leaves a rectangle in the same proportion to the original. The *silver ratio* describes a rectangle that, when *two* squares are removed, leaves a rectangle in the same proportion to the original. The *bronze ratio* does the same, but with *three* squares removed. The other metallic ratios proceed in the same way. These ratios have corresponding "Fibonacci-like" sequences. For example, the sequence for the silver ratio starts with 1 and 2, and the next number is generated as "twice the previous plus the penultimate," so that the sequence begins as 1, 2, 5, 12, 29, ... The sequence for the bronze ratio starts with 1 and 3, and the next number is "three times the previous plus the penultimate," so that the sequence begins 1, 3, 10, 33, 109, ... The other metallic ratios are defined similarly. The continued fractions are equally elegant. All of them have all 1s in the numerators, but whereas the golden ratio also has 1s down the diagonal, the silver ratio has all 2s, the bronze ratio has all 3s, and so on. It is often said that the golden ratio appears frequently in nature (seed patterns in sunflowers, galaxy shapes, etc.). It is more accurate to say that the *metallic* ratios appear often in nature—sometimes they are golden, and other times they are silver, bronze, etc.

CHAPTER 6
THE FUNDAMENTAL THEOREM OF ALGEBRA

ONLY A handful of mathematical theorems carry the moniker "fundamental." We have already encountered the Fundamental Theorem of Arithmetic. At the primary and secondary level, most students will only encounter one other: the Fundamental Theorem of Algebra.[118] As with the Fibonacci sequence, there is more than one way to tell the narrative of this theorem. I have opted to tell it as a story about the creation of numbers.

The German mathematician Leopold Kronecker (1823–1891) famously said, "God made the integers; all else is the work of man." I have two humble corrections. First, while the mathematical act is indeed a creative work on the part of the mathematician, it always remains the case that he is reaching for the heavens to bring down preexisting structures. But I don't mean to take anything away from the fact that the mathematician's creations are in some ways *his*. This tension is at the heart of the age-old question about whether mathematics is created or discovered. It seems to me that the resolution of this tension lies in the fact that the mathematician is a sort of "sub-creator." In his lecture-turned-essay "On Fairy-Stories," J.R.R. Tolkien refers to the author of fiction as a "sub-creator" and the bringing into being of fictional worlds, characters, and narrative arcs as an act of sub-creation. There is something similar going on when the mathematician invents a new structure. Second, if we are to attribute anything solely to the mind of God, it should probably be the natural numbers, not the integers. As we shall see, the negative numbers are not exactly natural. Thus, the story I am about to tell begins with a slight modification of Kronecker's premise: "God made the natural numbers," or if you prefer, "The natural numbers are the only natural ones."

Up until this point, we have been using the term *whole numbers* because the term itself described the property in which we were interested, i.e., "non-fractions" or even "numbers we use for counting objects." It is worth noting that many an elementary school math textbook, and sometimes those found in middle and high schools, places a lot of emphasis on the

118 Students who take calculus will come into contact with the Fundamental Theorem of Calculus. I thought of including this theorem in this section on important theorems with difficult proofs, but even stating the theorem itself, if one is to fully appreciate it, requires a bit of explanation. It was a painful cut, but likely for the best.

difference between natural numbers and whole numbers. The typical definitions are that natural numbers are the numbers 1, 2, 3, 4, 5, ..., whereas the whole numbers are the natural numbers together with 0, i.e., 0, 1, 2, 3, 4, 5, ... To be sure, the very nature of 0 as a number is categorically different than the other counting numbers. There is even an argument to be made that 0 does not really exist as a quantity, but only as the absence of quantity. After all, one cannot actually have zero apples; it is rather the case that one *does not have* apples. That said, professional mathematicians do not usually use the term *whole numbers*, opting for *natural numbers* instead. Further, in some college-level number theory texts, the term *natural numbers* means 0, 1, 2, 3, 4, 5, ..., whereas in others it means 1, 2, 3, 4, 5, ... Lest we think that mathematics is now teetering on the brink of relativism, this is really only a difference in terminology.[119] Whatever definition is chosen for a particular textbook, all of the theorems and explanations are phrased accordingly. Some will even avoid the whole thing and refer to the "positive integers" when they mean 1, 2, 3, 4, 5, ..., and the "non-negative integers" when they mean 0, 1, 2, 3, 4, 5, ...

For this chapter, I am going to adopt the term *natural numbers* to mean 0, 1, 2, 3, 4, 5, ... With apologies to those accustomed to the distinction in primary and secondary texts, I do this for three reasons. First, setting aside the aforementioned fascinating philosophical question about the nature of 0, I use the term *natural* because there *is* something natural about the numbers 0, 1, 2, 3, 4, 5, ... Small children learn these numbers early on, and that includes the number 0. At the very least, understanding what these numbers mean is a little easier than understanding what a fraction means, and significantly easier than understanding negative numbers. Second, I include 0 because there is something fundamental about this number, at least when it comes to thinking about addition. We mentioned before that the role of 0 in addition is like the role of 1 in multiplication. They are both *identities*, meaning that anything added to 0 is itself, just like anything multiplied by 1 is itself. While we certainly *can* think about addition without including 0, I much prefer to include it. Finally, given that we have adopted a way of writing numbers that

119 There is a similar disagreement about the definition of a trapezoid. Does a trapezoid have exactly one pair of parallel sides or at least one pair of parallel sides? It all comes down to whether or not you want rectangles and/or parallelograms to be trapezoids. The reality is that different works of mathematics have different definitions. This is not scandalous; it just means we must note carefully how a particular author is using the term and read the work consistently.

requires the use of 0—for example, in the number 1,708—it seems that there is something "basic" or "natural" about this digit.

With all that said, we start our story with the natural numbers and turn our attention to equation solving. Fear not; we are only interested in basic equations that use our two fundamental operations: addition and multiplication. Examples include equations such as $2x+3=9$, but also equations such as $2x^2+12=10x$. In the second equation, we are still only using addition and multiplication, but the term $2x^2$ includes multiplying x by itself. We call these types of equations *polynomial equations*, and I will use that term from this point on to avoid the potentially ambiguous term "basic equation."

Our quest is to find a universe of numbers that is "complete" in this way. That is, we would like to find a universe of numbers capable of solving equations that are composed using just these two operations. Note that this requirement is different than the universe being *closed* under addition and multiplication. After all, we already have closure even in the natural numbers under addition and multiplication, but as we shall see, we do not have universal equation-solving ability. Granted, *some* polynomial equations can indeed be solved. For example, the number 7 will solve the equation $2x+6=20$ because $2\times7+6=14+6$, which is 20.[120] The problem is that our universe of numbers is not capable of solving all polynomial equations. At the very least, the following basic equation has no solution.

$$x+1=0$$

Before you jump in and say, "Wait, isn't $x=-1$ a solution?" remember that our universe of numbers only has 0, 1, 2, 3, 4, 5, ... If this set is the entirety of numbers, then the above equation has no solution. Remember Dr. Tanton's advice? "If there is something in life you want, make it happen! (And deal with the consequences.)" Thus, if we do not have a

120 It is amazing, actually, that our training in equation solving has us apply algorithm after algorithm but rarely reminds us of what it is we are trying to do, which is to find values that make the statement *true*. If a student says, referring to the above equation, "The answer is 7 because $2 \times 7 + 6 = 20$," the response from the teacher should not be "It is great that you can do it in your head, but please show your work." Rather, the response should be "Well done, my good and faithful mathematician. You have not only found a solution but have proven your result. But why are you convinced that this is the *only* solution?" Alternatively, "Well done, my dear student. Try this one: $34{,}687 + 100{,}872x = 765{,}118$." The "mental" solution to the second problem will likely involve an unfolding of sorts that is akin to the desired algorithm. Algorithms do not exist to solve one problem. Many techniques can be applied to solve one problem. Algorithms exist to *conquer a category*. The real end of solving specific equations such as these is when students solve the general equation $ax + b = c$. When they do that, linear equations are conquered once and for all. *That* deserves to be written down.

solution to the equation, we *create one*. Again, lest mathematics at this point be accused of being relative, or still worse of dealing with things that are "not real," remember that this type of creative movement actually demonstrates the essence of mathematical sub-creation.[121]

Let us therefore invent a solution to $x+1=0$. With the power to create comes the power to name, and we will name our new number "−1." This indicates that the number is the "opposite of 1" because when added to 1 the result is nothing. Once we have this new number, we can explore how it acts with the other natural numbers. In other words, we are free to create, but not arbitrarily. In some ways we become bound by our mathematical creations because of the larger universe in which they are embedded.[122] For example, it will not take long to figure out that whenever -1 is added to some number, the answer is one less than that number. Multiplication requires a bit more thought. What is -1×2? We will not go through a full exploration of this, but once we start multiplying, we pick up new numbers: -2, -3, -4, -5, ..., etc. It turns out that each of the positive numbers is part of a pair involving a negative number, in which "pair" is understood as "adding together to get 0." (Zero has a pair as well: itself.) It also turns out that this extension of the natural numbers is "nice." In other words, all of the convenient properties of natural numbers—it doesn't matter in what order we add, the numbers are ordered so that any pair of non-equal numbers has a smaller one and a larger one, multiplication distributes over addition, etc.—still hold in this new universe, which we call the "integers."[123]

Of course, with the integers in place, we also get new polynomial equations such as $-3x+(-3)=6$. Many of these new equations have solutions in the original natural numbers, and others have solutions in the broader category of integers. The equation $-3x+(-3)=6$ has a solution of $x=-3$. The question at this point is "Have we completed our quest?" Do all

121 That said, this temptation to look at "−1" as somehow "not real" does point to something: it is not material. It is possible, however, that even the natural numbers are not material, that they exist only as immaterial ideals. I leave this question to the mathematical philosophers.

122 The same thing is true for the sub-creation that happens in a fairy tale. While the author can create fantastical worlds that defy even the laws of physics, this creativity is not without boundaries. Good must still be good, and evil must still be evil. In fact, in fairy tales, the malleability of certain things within the universe only serves to highlight those things that are not malleable, among which are the basic human condition and the nature of good and evil.

123 It is absolutely the case that there is a lot more to be done. The operations would need to be defined, properties would need to be proven, etc. Whole tomes have been written on the construction of various number systems and the continuity of properties.

polynomial equations in this new system have solutions? We do not need to look all that far to see that the answer is no. The problem comes from equations such as

$$2x = 1.$$

As before, we might be tempted to say, "Isn't the solution $x=\frac{1}{2}$?" Remember, though, our numerical universe is only $\ldots, -3, -2, -1, 0, 1, 2, 3, \ldots$, which doesn't include any number that when multiplied by 2 is equal to 1. But this equation is precisely what leads us to *invent* the number $\frac{1}{2}$.[124] As we did when bringing −1 into being, we now need to figure out how $\frac{1}{2}$ acts with the other numbers, and in particular what other numbers come into being along with $\frac{1}{2}$. In doing so, we get the rational numbers. And amazingly, it turns out that in gaining solutions to previously unsolvable equations, all the nice properties of integers are still true in the rational numbers. This is not a surprise. After all, our creations are conceived within an already ordered system of numbers, so they themselves bear the order of that system. Of course, along with new numbers, we also pick up new equations, e.g., $\frac{4}{9}x-1=0$. This one has a rational solution—can you find it? But is this always true? Is the universe of rational numbers complete? Do all polynomial equations built with rational numbers have solutions that are rational numbers? Unfortunately, the answer is still no. The equation below has no rational solutions, which we proved in the chapter on the existence of the irrational numbers.

$$x^2 = 2$$

The actual construction of the irrational numbers is difficult, so I will not present it here. The fact that they exist at all is enough for us. Nevertheless, once we combine the rational numbers with the irrational numbers, we have the full set of real numbers. We might think that our quest is over. After all, there is something about the real numbers that seems complete: if we draw a number line and label the points 0 and 1, we can select any point on the line and figure out the real number to which it corresponds, and we can take any real number and find its representative

124 We also saw that the rational numbers, and some irrational numbers, are "created" by drawing segments of particular lengths using a compass and a straightedge. There are numerous ways to think about what certain classes of numbers are, and there is no contradiction in presenting multiple ways of doing so in one book.

point on the line. It seems like the irrational numbers have "filled in" the rational numbers and given us a complete system.[125] Further, even some equations with irrational numbers in them have solutions in our newly expanded "real" universe of numbers, e.g., $3x-\sqrt{2}=\sqrt{5}$.

Given how complete the real numbers seem to be, it is perhaps a bit surprising to find that the following simple polynomial equation does not have any solution inside our new universe.

$$x^2=-1$$

It is easy to see why. A positive number times a positive number is always positive, a negative number times a negative number is also always positive, and 0 times 0 is 0. Therefore, we can never multiply a number by itself and get a negative number, so there is no real number that satisfies $x^2=-1$.

In looking for a solution to $x+1=0$, we created one and then extended our universe of numbers to the integers. In looking for a solution to $2x-1=0$, we created one and extended our universe to the rational numbers. Then we created a solution to $x^2=2$ and extended our universe to include all real numbers. Why not do the same here? Why not invent a solution to $x^2=-1$? In creating such a number, we have the privilege of naming it, and we will call this solution i.

It is an unfortunate reality of nomenclature that the use of the letter i comes from the term "imaginary," and that numbers such as these are known as *imaginary numbers*. It is not incorrect, mind you. "Imaginary" does mean in some way "not real," and these numbers are not in the official set of "real numbers." Further, "imaginary" also means "made up," and indeed we did just "make up" the number i. But they are no more "unreal" and no more "made up" than the negative numbers. All along the way, we have created numbers, extended our universe, and watched amazing things happen. The fact that these numbers are termed "imaginary" often gives license to think that they are not worth our study. They are, and as we will see, they bring about a surprising result.

125 I am skipping for now the distinction between rational numbers and algebraic numbers. We will introduce the idea briefly in the next chapter. It is true, however, that the creation of a new set after the rational numbers leads only to the algebraic numbers, not the full set of real numbers. This bit of hand-waving is acceptable for now; it does not significantly change the narrative leading to the Fundamental Theorem of Algebra.

For clarity, I should note that the numbers we are really interested in are more properly called the *complex numbers*. In other words, we are concerned with not only adding i into the mix of real numbers but also combining it with all of the real numbers through addition and multiplication. (This scenario is similar to how the introduction of the number -1 led to the creation of $-2, -3, -4, -5, \ldots$ through multiplication.) The complex numbers include things like 3, $\frac{4}{9}$, -5, and $\sqrt{7}$, as well as i, but they also include things like $3i$, $3+(-5)i$, $\sqrt{7}+\frac{4}{9}i$, etc.

When we construct the complex numbers, something remarkable happens. We get a solution to not only $x^2=-1$ but also all polynomial equations that use real numbers. (Sometimes we get more than one solution. In fact, both i and $-i$ are solutions to $x^2=-1$.)

The Not-Quite-Fundamental Theorem of Algebra
Any polynomial equation put together with real numbers will have at least one solution in the complex numbers.

(Note that the solution guaranteed by the theorem actually might be real, as in the solution to $x+4=7$. Remember, though, that the complex numbers *contain* the real numbers. Therefore, such equations do not contradict the theorem.)

This theorem is rather incredible. No matter how complicated our polynomial equation, it may not have a *real number* solution, but it will at least have a *complex number* solution. This includes equations such as

$$4x^{15} + \sqrt{13}x^6 + (-2)x^3 + \tfrac{11}{17}x + 19 = 0.$$

It may be hard to find—it may be hard to even write down—but it has a solution. The dilemma, though, is that the introduction of the complex numbers also introduces new equations such as

$$(4+i)x^5 + (2+3i)x^2 + 7x + 2 = 0.$$

Here we arrive at the surprising happy ending to our fairy tale. Equations such as the one above also have solutions, *and those solutions are in the complex numbers*. In other words, as we journeyed through our acts of mathematical sub-creation, we might have started to get the impression that new numbers might solve old equations, but they also give us new equations, some of which might not have solutions, so we introduce new numbers, and so on. We might think that this cycle is endless, but it turns out that it is not. The process stops here. Even though introducing complex numbers does introduce new equations, those equations will still have solutions in the complex numbers, which is precisely the Fundamental Theorem of Algebra.[126]

The Fundamental Theorem of Algebra
Any polynomial equation assembled with complex numbers will have at least one solution in the complex numbers.

Amazingly, our journey is complete. All polynomial equations—those formed by using only the two basic operations of addition and multiplication—will have a solution in the complex numbers. In fact, most will have more than one solution. In this quest, there is no need to create any more numbers.[127] In some weird way, this makes the complex numbers more "real" than the incomplete real numbers.

126 The Fundamental Theorem of Algebra actually says that any non-constant polynomial with complex coefficients will always have a complex root. A *root* is simply a value that makes the polynomial expression equal to 0. It was easier to avoid that discussion here. The two formulations are equivalent. Moreover, we can predict the number of solutions based on the largest exponent of x. If the largest exponent is 6, there are exactly six solutions, though this involves understanding that some solutions count "more than once." The formulation that "a polynomial with complex coefficients and degree n has exactly n roots counting multiplicity" is sometimes stated as the Fundamental Theorem of Algebra. This is also equivalent to the formulation we have chosen here.

127 Mathematicians *have* invented numbers beyond the complex numbers. Some of these include the quaternions, octonions, etc., often termed *hypercomplex numbers*.

CHAPTER 7
EULER'S IDENTITY

IN THE first chapter of this part of the book, we were introduced to the number π, and we saw that it was an *irrational* number, meaning we cannot write it as a fraction of whole numbers. In the previous chapter, we were introduced to the number i, which by definition acts as a square root to -1 but also serves a critical role in creating a system of numbers capable of solving all basic equations formed by addition and multiplication. There is no doubt that these are two of the most important numbers in all of mathematics. We have also made the argument a few times that the numbers 0 and 1 have a privileged place, at least as they relate to the basic operations of addition and multiplication. In this chapter, we present a fifth number that is perhaps equally as important as π.

One of the first references to this fifth number appears in the work of the Scottish mathematician John Napier (1550–1617), who is sometimes referred to as "Marvellous Merchiston," Merchiston being his birthplace. But the number only appears buried in a table of values Napier was calculating, and it is unclear how much he knew about it and where it comes from. Its discovery is attributed to the later work of Jacob Bernoulli. Bernoulli discovered the number by examining monetary interest. He started by looking at a $1 investment that earns 100% interest every year. (How many of us would love to have an account paying such an interest rate!) After one year, the account earns 100% of $1, or $1 in interest. This means that there will be $2 (the original $1 plus the $1 of interest) in it at the end of the year.

But what if we split the 100% interest and award 50% at the end of six months and the other 50% at the end of the year? In six months, the account earns 50% of $1, or $0.50 in interest, leaving $1.50 ($1 + $0.50) in the account. Here is where Bernoulli's account gets interesting.[128] At the end of the second half year, the account earns 50% of the $1.50 that has been sitting there for six months. This amounts to 0.50 × $1.50, or $0.75 in interest, and leaves at the end of the year $1.50 + $0.75 = $2.25 in the account. This amount is *more* than $2! Why did this happen? The simple answer is that the interest for the second half year was calculated not only on the original amount but also on the interest earned in the first

128 Pun intended.

half year. In other words, we earned *interest on the interest*. In fact, the extra \$0.25 is explained *precisely* by the interest on the interest. The interest earned during the first half year was \$0.50, and 50% of \$0.50 is \$0.25. This is the phenomenon of compound interest.

What if we compound the interest every three months? In other words, what if we opt to take the 100% interest rate in installments of 25% four times during the year? The following chart calculates how much is in the account at the end of one year:

Quarter	Initial Amount	Interest	Ending Amount
I	\$1.00	25% of \$1.00 = \$0.25	\$1.00 + \$0.25 = \$1.25
II	\$1.25	25% of \$1.25 = \$0.3125	\$1.25 + \$0.3125 = \$1.5625
III	\$1.5625	25% of \$1.5625 = \$0.390625	\$1.5625 + \$0.390625 = \$1.953125
IV	\$1.953125	25% of \$1.953125 = \$0.48828125	\$1.953125 + \$0.48828125 = \$2.44140625

Of course, in real money, the amount at the end is rounded to \$2.44. The point, however, is that this amount is greater than \$2.25 (when the interest was split over two six-month periods), which itself was greater than \$2.00 (when the interest was only awarded once at the end of the year). What would happen if we split the interest every month, every day, every hour, or even every second? Can we just keep making more and more money? On the one hand, yes. The value of the money in the account at the end of the year does increase every time we increase the number of times the interest is compounded, but the fascinating thing is that the account balance does not increase *without bound*. In other words, there is a limit to how much interest the account can earn. That limit occurs when we compound the interest "at every moment," and the value turns out to be about \$2.718281828459....[129] Bernoulli knew that this "limit" existed and knew its approximate value. What he was not able to prove is that this number, like π, is irrational. It cannot be written as a fraction of

129 Sometimes e is known as the "Andrew Jackson" number. Andrew Jackson served two terms, was the seventh president of the United States, and was elected in 1828. The number e is approximately 2.718281828.... Remember, however, that the "1828" does not keep repeating. If it did, e would be rational.

whole numbers. That feat was accomplished by none other than Leonard Euler (who was a student of Bernoulli's younger brother Johann).

The letter we use for this number is e, and it is often called "Euler's number." There are any number of ways to define e, though there is no widespread agreement on which one is *the* definition. This does not mean that mathematicians disagree on the *value* of e, just that some texts will give one of several definitions and present the others as properties of e. It would be like wanting to *define* π as $4-\frac{4}{3}+\frac{4}{5}-\frac{4}{7}+\frac{4}{9}-\ldots$ (a property we saw earlier), and then proving a theorem that states, "The ratio of the circumference to the diameter in any circle is always π." This would not be mathematically incorrect, but it would defy historical convention, which starts with defining π in terms of circles and proceeds from there. With e it is different. Although we know several expressions that equal e and several ways of characterizing it, there is no consensus on which one to choose as a definition. Any number of ways can serve that purpose, and then it would be the task of the mathematician to prove the others as properties.

For historical reasons, we will define e in terms of compound interest.

> DEFINITION: The number e is the amount of money in an account after one year if the account started with \$1.00 and earned 100% interest that is compounded continuously.

There is a virtually identical way of defining e.[130]

> ALTERNATE DEFINITION: The number e is the value of $\left(1+\frac{1}{n}\right)^n$ as n gets larger and larger.

I will only summarize the reason why these two definitions are equivalent. The argument begins by looking at compounding interest "n times per year." The interest at 100% for each "compounding period" is $\frac{100\%}{n}$, or $\frac{1}{n}$. The amount of money in the account after this interest installment is \$1.00 $+\frac{100\%}{n}$ or $1+\frac{1}{n}$. This process is repeated for n installments. I will leave it to you to figure out the details.

Like with π, there is work to be done to show that e is a number at all.

130 Written with limits, e is $\lim_{n\to\infty}\left(1+\frac{1}{n}\right)^n$.

In other words, how did Bernoulli know that the increasing number of "compounding periods" will result in a number that is not infinity? That question is a hard one and outside the scope of this book. Nevertheless, the process of compounding interest does have a finite limit, so e is in fact an actual number.

Once we adopt Bernoulli's definition, some remarkable properties of e arise. Like π, the most famous one has to do with an infinite sum. This is so famous that it is sometimes used as the definition of e.[131]

$$e = 1 + \frac{1}{1} + \frac{1}{2\times 1} + \frac{1}{3\times 2\times 1} + \frac{1}{4\times 3\times 2\times 1} + \frac{1}{5\times 4\times 3\times 2\times 1} + \cdots$$

There are three infinite continued fractions that are remarkably patterned and equal to e. One consists of the positive integers in ascending order occurring in both the numerators and the diagonal.

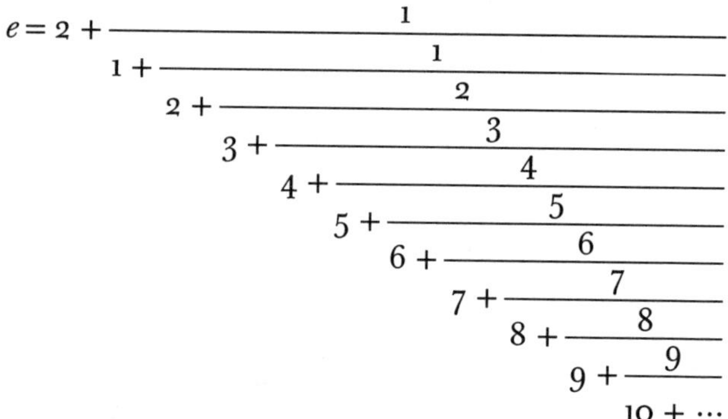

$$e = 2 + \cfrac{1}{1 + \cfrac{1}{2 + \cfrac{2}{3 + \cfrac{3}{4 + \cfrac{4}{5 + \cfrac{5}{6 + \cfrac{6}{7 + \cfrac{7}{8 + \cfrac{8}{9 + \cfrac{9}{10 + \cdots}}}}}}}}}}$$

A second one has all 1s in the numerator.

131 Using factorial notation, this expression can be written as $e=\frac{1}{0!}+\frac{1}{1!}+\frac{1}{2!}+\frac{1}{3!}+\cdots$, where 0! and 1! are understood to be 1.

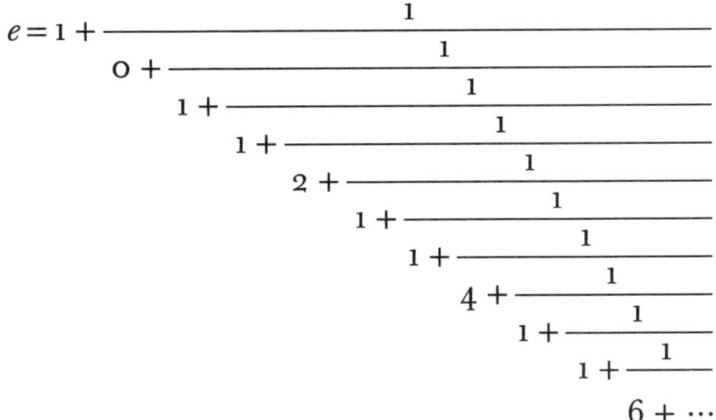

The numbers down the diagonal have a pattern when grouped by threes: (1, 0, 1), (1, 2, 1), (1, 4, 1), (1, 6, 1), (1, 8, 1), ... The middle numbers in each triplet are the even numbers starting at 0.

A third continued fraction has a similar list of numerators, with only the first one being a 2.

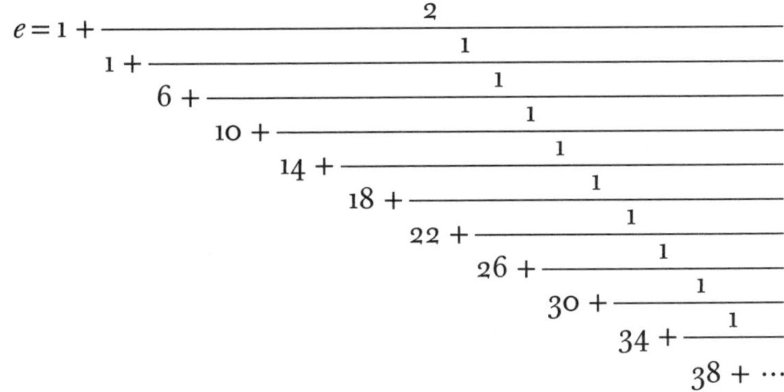

The numbers down this diagonal are 1, 1, 6, 10, 14, 18, 22, 26, ... After the first two 1s, the others increase by 4 every time.[132]

Finally, there are two rather interesting probability scenarios using the number $\frac{1}{e}$.

1. Suppose someone plays a slot machine that pays out with a probability of 1 in n and plays it n times. The probability of losing *every time* is, for large values of n,

132 One additional "definition" of e that is quite common depends on one's understanding about graphs and tangent lines. The number e is the unique value such that the slope of $y = e^x$ at the point x is e^x. In other words, it is the unique value such that the slope of the tangent line at any given point is equal to the y-coordinate at that point.

approximately $\frac{1}{e}$. (Bernoulli came up with this scenario.)

2. There are n people checking their hats at a party. The person checking hats does not keep track of them and randomly hands them out at the end of the night. The probability that *no* hat gets back to the rightful owner is, for large values of n, approximately $\frac{1}{e}$.

It turns out that e and π are representatives of a particular class of numbers that is perhaps even more interesting than irrational numbers or constructible numbers. This category has a lot to do with the Fundamental Theorem of Algebra.

> DEFINITION: If a number solves a polynomial equation built only with integers, it is an *algebraic number*. A number that is not algebraic is called a *transcendental number*.

The number 3 is algebraic because it is the solution to the equation $4x+8=20$. The rational number $-\frac{4}{5}$ is also algebraic because it solves the equation $-5x=4$. In fact, all rational numbers are algebraic, and therefore so are all integers. There are, however, some irrational numbers that are also algebraic. The irrational number $\sqrt{2}$ is algebraic because it solves the equation $x^2-2=0$.

Another way to think about an algebraic number is to start with the number and apply addition, subtraction, multiplication, division, and exponentiation (which is really just repeated multiplication) using only natural numbers in an attempt to arrive at 0. For example, 3 is algebraic because we can subtract 3 and get 0.[133] (We can also do strange things like multiply by 2, add 1, then subtract 7, but finding one way of getting to 0 is enough.) The number $-\frac{4}{5}$ is algebraic because we can multiply by 5 and then add 4 to get 0. The number $\sqrt{2}$ is algebraic because we can square it and then subtract 2. Even numbers such as $\sqrt{2}+\sqrt{3}$ are algebraic, but I will leave it to you to figure out how to get to 0.

In fact, all *constructible* numbers are algebraic. Remember that a

133 I have the Numberphile video "Transcendental Numbers" (https://www.youtube.com/watch?v=seUU2bZtfgM) to thank for the "bringing a number back down to 0" way of thinking about algebraic numbers. This way of thinking is much simpler to comprehend than "solving a polynomial equation" and shows more clearly what it is that the transcendental numbers transcend.

constructible number is the length of a segment that can be drawn starting with a segment of length 1 and using only a straightedge and a compass. The fact that these are all algebraic may not be a surprise, though. Recall that the constructible numbers can also be created with the integers, the four basic operations, and square roots. It makes sense that these numbers can then be "undone" using addition, subtraction, multiplication, division, and squaring. There are, of course, numbers that are not constructible but are still algebraic. For example, we know from our discussion of impossible constructions that $\sqrt[3]{2}$ is non-constructible. It is algebraic, though. It is a solution to the equation $x^3 = 2$, or, alternatively, we can get from $\sqrt[3]{2}$ to 0 by cubing the number and subtracting 2.

It seems that *most* numbers we have talked about are algebraic. What is incredible about π and e is that they are *not* algebraic, and therefore are transcendental. The name is not accidental. They are not the solution to any polynomial equation that uses integers, and they cannot be "brought down to 0" using the natural numbers, addition, subtraction, multiplication, division, and exponentiation. They *transcend* the other numbers and the basics of arithmetic. Each transcendental number stands alone in all of its strange and mysterious glory. The following diagram helps us to understand how this category is situated within the positive real numbers.[134]

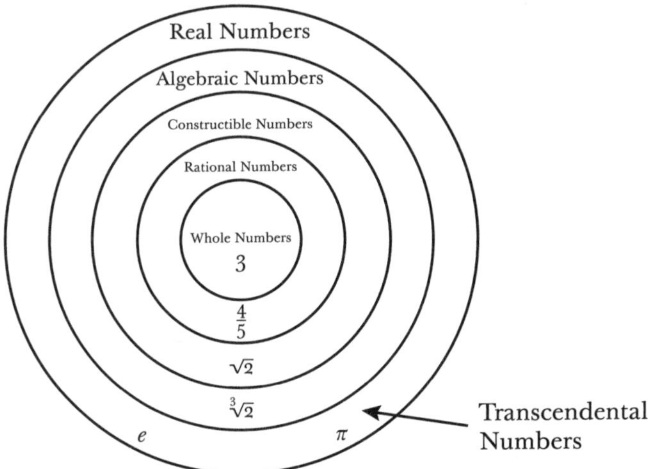

134 The diagram with negative numbers and complex numbers is harder to draw. The number i is algebraic but not constructible. The number -3 is rational, and most consider it constructible only because the constructible numbers are those that can be constructed by a compass and a straightedge *and their negatives*. That is a bit of a strange definition, which I have opted to put to the side for the sake of this book.

The scale of the diagram is inaccurate: the size of the set of transcendental numbers is *huge* compared with the algebraic numbers. In the chapter on counting the irrational numbers, we noted that "nearly all" of the real numbers are irrational. It turns out that it is precisely the transcendental numbers that make the set of irrational numbers so large. *Most* real numbers are transcendental. Equally true is the fact that *we do not know very many transcendental numbers*. We have two examples in π and e, and mathematicians certainly know of more than just these two, but the list is fairly small.

Being rare representatives of this mysterious category of numbers may lend even more credibility to the heroic nature of π and e. Together with 0, 1, and i, these five numbers are widely considered to be the most important numbers in all of mathematics. It turns out that these five numbers are related to each other in an elegant formula called Euler's Identity. The history of the formula is a bit muddy. It seems that some mathematicians may have known of it prior to Euler, but it nevertheless retains his namesake. The proof[135] is difficult and involves understanding how to apply exponents that are complex numbers. Even so, the result is of monumental importance and displays a beautiful simplicity. At least in this chapter, it deserves to have the last word.

Euler's Identity

$$e^{\pi i}+1=0$$

135 There are several ways to prove Euler's Identity. One way starts with the expansion of e that we saw in this chapter, $e=\frac{1}{0!}+\frac{1}{1!}+\frac{1}{2!}+\frac{1}{3!}+\cdots$, and generalizes it to $e^x=1+\frac{x}{1!}+\frac{x^2}{2!}+\frac{x^3}{3!}+\ldots$. The second part of the proof involves two similar facts about trigonometric functions: $\sin(x)=\frac{x}{1!}-\frac{x^3}{3!}+\frac{x^5}{5!}-\frac{x^7}{7!}+\cdots$ and $\cos(x)=1-\frac{x^2}{2!}+\frac{x^4}{4!}-\frac{x^6}{6!}+\cdots$. With these three expansions, you can substitute πi into the formula for e^x, substitute π into the formulas for $\sin(x)$ and $\cos(x)$, multiply the formula for $\sin(x)$ by i, and then simplify using the fact that $i^2=-1$ to see that $e^{\pi i}=\cos(\pi)+i\sin(\pi)$. Finally, if you know some trigonometric values, you can simplify this equation to $e^{\pi i}=-1+i(0)$, or $e^{\pi i}=-1$. Add 1 to both sides of the equation, and you are done. It is noteworthy that you can actually substitute ix instead of $i\pi$ and perform the same algebra to get Euler's Formula, which is $e^{ix}=\cos(x)+i\sin(x)$. Euler's Formula is particularly helpful because it now allows us to understand and apply complex number exponents. The full version of this proof would involve arguing the above expansions, which come from the Taylor series, often found in a Calculus II course.

CHAPTER 8
THE PRIME NUMBER THEOREM

WE HAVE already seen the astonishing fact that there are an infinite number of prime numbers. But the way they appear seems to be without pattern, or *almost* without pattern. A minor pattern we see is a plethora of small primes (2, 3, 5, 7, 11, 13, 17, 19, etc.), whereas it tends to be harder to find larger primes. In other words, the primes seem to spread out as they get larger. This all *seems* to be true, anyway—but are prime numbers really denser among small numbers? In other words, do they become more spread out as the numbers increase? One reason to think this is true is because the larger a number is, the more possibilities it has for divisors. In fact, there is a way we can generate prime numbers that supports this progressive spreading out. This process is known as the Sieve of Eratosthenes, attributed to the ancient Greek mathematician from Cyrene.[136]

To begin, write down the sequence of numbers in which you want to identify the primes. For now, we will use 50 as our maximum. In other words, our goal is to generate all of the prime numbers from 2 through 50. Notice that we do not list 1, because *1 is not a prime number!*

	2	3	4	5	6	7	8	9	10
11	12	13	14	15	16	17	18	19	20
21	22	23	24	25	26	27	28	29	30
31	32	33	34	35	36	37	38	39	40
41	42	43	44	45	46	47	48	49	50

136 Eratosthenes was much more than a mathematician. He also studied music, geography, and astronomy, and even wrote poetry. He was a true "Renaissance Man" long before the Renaissance.

Start by eliminating all multiples of 2 except for 2 itself. After all, any multiple of 2 that is greater than 2 cannot be prime: it has a factor of 2.

	2	3		5		7		9	
11		13		15		17		19	
21		23		25		27		29	
31		33		35		37		39	
41		43		45		47		49	

Next, eliminate all multiples of 3 except for 3 itself. These cannot be prime, either, because they are divisible by 3.

	2	3		5		7			
11		13				17		19	
		23		25				29	
31				35		37			
41		43				47		49	

Continue this process by finding the next number that hasn't been hit and eliminate its multiples (except for itself). For example, notice that we do not have to worry about multiples of 4 because there are none left: they were taken care of when we eliminated multiples of 2. Therefore, the next number to consider is 5. Eliminate all multiples of 5 except for 5 itself.

	2	3		5		7			
11		13				17		19	
		23						29	
31						37			
41		43				47		49	

We do not have to worry about the number 6 and its multiples. These first got hit when eliminating multiples of 2, and they would have been hit again when eliminating multiples of 3. Either way, they are no longer in our table. The next number still in the table is 7, so we eliminate its multiples (except itself).

	2	3		5		7			
11		13				17		19	
		23						29	
31						37			
41		43				47			

The fact that the sieve produces a list of primes is not the genius of the algorithm. After all, we simply crossed out numbers that have proper divisors. By definition, only the prime numbers will be left. In other words, the crude summary of the algorithm is "If you want to generate a list of prime numbers, first list all numbers, then systematically cross out the ones that are not prime." Stated this way, it seems like common sense, not genius. The real genius of the algorithm is found in the *speed* with which it operates. For the list above, notice that we are actually finished. After only *four* elimination passes, only prime numbers are left: 2, 3, 5, 7, 11, 13, 17, 19, 23, 29, 31, 37, 41, 43, and 47. How do we know we can stop? Well, we certainly would never have to deal with crossing off multiples of any number greater than $50 \div 2 = 25$. For example, there can be no multiples of 26 that are less than 50. (Think about why this is true.) But we don't even need to go all the way to 25. Our list was complete after taking care of multiples of 7! Why? For example, why are we sure that there are no multiples of 11 left in our list? After all, unlike 26, the number 11 *does* have multiples that are less than 50: 22, 33, and 44. Notice, however, that all of these have already been crossed off the list. We could have continued our sieving process, but we would not have eliminated any other numbers. I leave it to you to think about why stopping after 7 was sufficient for primes under 50 and what the magic stopping number is for the general case. For example, how far must we go if we want all primes less than or equal to 100?[137] What about less than or equal to n?

The Sieve of Eratosthenes supports the observation that the prime numbers spread out as they increase. It is reasonable to assume that when performing these eliminations, larger numbers have a higher probability of being hit, so they end up not being prime. For example, the number 7 could only have been crossed out had it been hit by being a multiple of 2, 3, or 5, but the number 47 had to survive being a multiple of 2, 3, 5, *and* 7. Numbers like 1,001 would have to survive even more rounds of elimination. With more possibilities for being hit, there seems to be a smaller likelihood of survival.

In addition to the Sieve of Eratosthenes, there is a known theorem that supports the spreading out of the primes.

> *Consecutive Non-Primes Theorem*
> We can find arbitrarily large gaps in the primes.

For example, if we want to find a sequence of ten numbers in a row containing no prime numbers at all, we can do that. How might we accomplish this? We might use the Sieve of Eratosthenes repeatedly to look for a gap this large, but we don't really know how long that might take. The biggest gap in the first twenty-five primes is between 89 and 97, which

137 I don't want to give away the general principle here, but I will reveal that stopping after multiples of 7 is also sufficient for generating primes under 100, but not sufficient if you want primes under 150. The number 143 is not prime but needs to be crossed out when considering multiples of 11. I mentioned above that the impressiveness of the Sieve of Eratosthenes is found in the fact not only that it works but that it works quickly. To find the primes less than 1,000,000, you only have to cross out multiples of numbers less than 1,000.

has only seven consecutive non-primes in this gap: 90, 91, 92, 93, 94, 95, and 96. Rather than running this algorithm over and over, hoping for the best, we would prefer to explicitly *construct* ten consecutive composite numbers. One way to do this is to start with the following number.

$$(11 \times 10 \times 9 \times 8 \times 7 \times 6 \times 5 \times 4 \times 3 \times 2) + 2$$

We should first convince ourselves that this number cannot be prime. To do that, we need a factor, which in this case is 2. How do we know that this number is divisible by 2? Let's look at the first part of the number: $11 \times 10 \times 9 \times 8 \times 7 \times 6 \times 5 \times 4 \times 3 \times 2$. This part *explicitly has a factor of 2 in it!* Next, observe that multiples of 2 are exactly two apart. Therefore, if we start with a multiple of 2, e.g., $11 \times 10 \times 9 \times 8 \times 7 \times 6 \times 5 \times 4 \times 3 \times 2$, and then add 2 to it, we have *another* multiple of 2. Therefore, because $11 \times 10 \times 9 \times 8 \times 7 \times 6 \times 5 \times 4 \times 3 \times 2$ has a factor of 2, $(11 \times 10 \times 9 \times 8 \times 7 \times 6 \times 5 \times 4 \times 3 \times 2) + 2$ must also have a factor of 2.

Okay, so we have a non-prime—but we are looking for *ten* non-primes in a row. The next number after $(11 \times 10 \times 9 \times 8 \times 7 \times 6 \times 5 \times 4 \times 3 \times 2) + 2$ is

$$(11 \times 10 \times 9 \times 8 \times 7 \times 6 \times 5 \times 4 \times 3 \times 2) + 3.$$

Can you see by a similar argument that this number is also not prime? The first part, $11 \times 10 \times 9 \times 8 \times 7 \times 6 \times 5 \times 4 \times 3 \times 2$, explicitly has a factor of 3, and multiples of 3 are exactly three apart. Therefore, if $11 \times 10 \times 9 \times 8 \times 7 \times 6 \times 5 \times 4 \times 3 \times 2$ is a multiple of 3, so is $(11 \times 10 \times 9 \times 8 \times 7 \times 6 \times 5 \times 4 \times 3 \times 2) + 3$. Therefore, with 3 as a factor, this number cannot be prime.

It is not difficult to see a path forward. The following numbers represent the sought-after ten composite numbers in a row.

Number	Factor that shows it is not prime
$(11\times10\times9\times8\times7\times6\times5\times4\times3\times2)+2$	2
$(11\times10\times9\times8\times7\times6\times5\times4\times3\times2)+3$	3
$(11\times10\times9\times8\times7\times6\times5\times4\times3\times2)+4$	4
$(11\times10\times9\times8\times7\times6\times5\times4\times3\times2)+5$	5
$(11\times10\times9\times8\times7\times6\times5\times4\times3\times2)+6$	6
$(11\times10\times9\times8\times7\times6\times5\times4\times3\times2)+7$	7
$(11\times10\times9\times8\times7\times6\times5\times4\times3\times2)+8$	8
$(11\times10\times9\times8\times7\times6\times5\times4\times3\times2)+9$	9
$(11\times10\times9\times8\times7\times6\times5\times4\times3\times2)+10$	10
$(11\times10\times9\times8\times7\times6\times5\times4\times3\times2)+11$	11

To be sure, there may be sequences of ten consecutive composite numbers that occur *sooner* than these, but that doesn't matter for our purposes. What matters is that we can find at least one gap of ten in the primes.[138]

What if we wanted to find a gap of one hundred consecutive numbers without a prime? We could follow the same procedure. I leave it to you to verify that there are indeed one hundred numbers in this list.

Number[139]	Factor that shows it is not prime
$(101\times100\times99\times98\times97\times96\times95\times\ldots\times6\times5\times4\times3\times2)+2$	2
$(101\times100\times99\times98\times97\times96\times95\times\ldots\times6\times5\times4\times3\times2)+3$	3
$(101\times100\times99\times98\times97\times96\times95\times\ldots\times6\times5\times4\times3\times2)+4$	4
$(101\times100\times99\times98\times97\times96\times95\times\ldots\times6\times5\times4\times3\times2)+5$	5
...	...
$(101\times100\times99\times98\times97\times96\times95\times\ldots\times6\times5\times4\times3\times2)+99$	99
$(101\times100\times99\times98\times97\times96\times95\times\ldots\times6\times5\times4\times3\times2)+100$	100
$(101\times100\times99\times98\times97\times96\times95\times\ldots\times6\times5\times4\times3\times2)+101$	101

Of course, there was no particular reason we chose ten or one hundred.

138 Following the number 113 we find thirteen composite numbers in a row, certainly satisfying our search for a gap of ten.

139 If we wanted to use the factorial symbol to write the numbers in this table, they would be 101! + 2, 101! + 3, 101! + 4, 101! + 5, ..., 101! + 99, 101! + 100, 101! + 101. Using this factorial symbol, a proof of the formal case for the number *n* would be cleaner.

We could use this same technique to find gaps in the primes that are *as large as we want*, and this process essentially proves the theorem. The theorem hints that the primes seem to spread out the larger they become because we can find arbitrarily large gaps of numbers between them.

Allow me to add one more fact to the mounting evidence supporting the spreading out of the prime numbers as they increase. There is a particularly important and famous theorem that specifically measures this spread. The theorem actually approximates how many primes we can expect within a certain range of numbers. I am going to state this theorem in its "official" version first. A fair warning, however: the theorem contains a logarithm. Because I will not go into the definition of a logarithm, I will also state the theorem in some easier-to-understand language.

> *Prime Number Theorem*
> The number of primes less than or equal to n is approximately $\frac{n}{\ln(n)}$. This approximation becomes more accurate as n increases, so that for very large values of n there is very little difference between $\frac{n}{\ln(n)}$ and the number of primes less than n.

We can restate this theorem in terms of the number of digits of n.[140]

> *Restated Prime Number Theorem*
> The number of primes less than or equal to n is approximately (0.4342944819... × n) ÷ (the number of digits in n).
> This approximation becomes more accurate as n increases.

As an example, let's look at how many prime numbers we would expect to find between 1 and 999,999. The Restated Prime Number Theorem says that we should expect

$$(0.4342944819\ldots \times 999{,}999) \div (\text{the number of digits in } 999{,}999).$$

140 This restatement is not entirely accurate. The restatement should say $\frac{n}{\ln(10) \times \text{the number of digits in } n}$. I opted for a decimal approximation to avoid logarithms. Moreover, even though these are equivalent statements, it is the case that $\frac{n}{\ln(n)}$ provides better approximations than $\frac{n}{\ln(10) \times \text{the number of digits in } n}$ does. That said, mathematicians have some more advanced formulas that are better at approximating the number of primes than even $\frac{n}{\ln(n)}$. But $\frac{n}{\ln(n)}$ is really good, and it is really elegant. Whether or not the complicated proof is in *The Book*, the theorem itself most certainly is.

This expression is the same as 434,294.0476... ÷ 6, or approximately 72,382. The number of primes between 1 and 999,999 is actually 78,498. How close is 72,382 to 78,498? It may not seem that close, but considering how big these numbers are, it is pretty good. Moreover, these estimates become more accurate as *n* increases. For example, if we look at the number of primes less than 999,999,999,999,999, our approximation formula gives us 28,952,965,460,217, and the actual number of primes is 29,844,570,422,669. For numbers this large, the approximation is very good, especially if we think about it in terms of the percentage of numbers less than 999,999,999,999,999 that are prime.[141] That is the other use of the Prime Number Theorem: to estimate the percentage of numbers within a certain range that are prime.

n	Approximate percentage of the first n numbers that are prime
100	21.7%
1,000	14.5%
10,000	10.9%
100,000	8.7%
1,000,000	7.2%
10,000,000	6.2%
100,000,000	5.4%
1,000,000,000	4.8%

Clearly the primes become less common, or more spread out, as numbers increase. There is also a consequence of the theorem describing how the likelihood of a number being prime is related to the number of digits.

141 Another way to think about how "close" the two numbers are to one another is by "percent error." In the case of 72,382 versus the actual count of 78,498, the percent error is about 7.8%. In the case of 28,952,965,460,217 versus the actual number 29,844,570,422,669, the percent error is less than 3%. This percent error will approach 0% as *n* increases.

A Consequence of the Prime Number Theorem
A number with d digits is approximately twice as likely to be prime as a number with $2 \times d$ digits.

In other words, the probability of a 50-digit number being prime (approximately 0.87%) is about twice the probability of a 100-digit number being prime (approximately 0.43%). This relationship based on the number of digits is the result of the logarithm formulas.[142]

The Prime Number Theorem is extremely difficult to prove. (In fact, it is one of the harder theorems in this book to even state and understand.) Nevertheless, it is one of the most beautiful and important theorems in all of number theory. It gives a hint about how to predict the occurrence of these mysterious prime numbers that seem to occur without pattern. It at least tells us that, on average, they seem to become more spread out as numbers increase. That said, it is an almost unbelievable coincidence that the numbers 18,409,199 and 18,409,201 are both prime, as are 6,000,000,000,581 and 6,000,000,000,583. The numbers in each pair are only two apart! Thus, even though the primes continue to spread further and further apart as a general rule, we continue to find examples that are shockingly close together.

142 The probability of a number no greater than n being prime is approximately $\frac{1}{\ln(n)}$, which also means that the average gap between primes for numbers no greater than n is $\ln(n)$. So, for example, if you choose a random number less than one trillion (1,000,000,000,000), the probability of its being prime is about $\frac{1}{\ln(1{,}000{,}000{,}000{,}000)}$, or somewhere around $\frac{1}{27}$ or $\frac{1}{28}$, which is around 3.6%. This also means that the average gap between primes less than one trillion is about 27 or 28, which is approximately $\ln(1{,}000{,}000{,}000{,}000)$.

CHAPTER 9
FERMAT'S LAST THEOREM

WE RETURN once more to one of the most famous theorems in all of mathematics: the Pythagorean Theorem.

The Pythagorean Theorem
In any right triangle, the square of the hypotenuse is equal to the sum of the squares of the two legs.

In other words, $a^2+b^2=c^2$, where a and b are the lengths of the two legs in a right triangle and c is the length of the hypotenuse. As side lengths, a, b, and c can be any positive number, including fractions and irrational numbers. If the lengths are whole numbers, then the trio is called a *Pythagorean triple*. For example, if $a=3$, $b=4$, and $c=5$, then $3^2+4^2=5^2$. The 3-4-5 triangle is the world's most famous right triangle.

It is quite easy to generate infinitely many Pythagorean triples. For example, we can take the 3-4-5 triangle and multiply each side by 2 to get 6-8-10. You can check the equation to make sure that it is true. In fact, we could have multiplied 3, 4, and 5 by any number we wanted. Let's try multiplying by 13 and then check the Pythagorean relationship.

$$39^2+52^2=1{,}521+2{,}704=4{,}225$$
and
$$65^2 = 4{,}255$$

Why did this work? Why can we multiply 3, 4, and 5 by whatever number we want and still get a Pythagorean triple? There are two ways to see this. Algebraically, if we multiply 3, 4, and 5 by an arbitrary number n, we get $3n$, $4n$, and $5n$. We then square each side.

$$(3n)^2=3n\times 3n=9n^2$$
$$(4n)^2=4n\times 4n=16n^2$$
$$(5n)^2=5n\times 5n=25n^2$$

We can now check the Pythagorean relationships: $9n^2 + 16n^2 = 25n^2$. Not only is this equation true, but it is true precisely *because* of the underlying 3-4-5 triple: $3^2 + 4^2 = 5^2$.

We can also verify the property geometrically. Remember that the 3, 4, and 5 were not just numbers; they were sides of a right triangle.

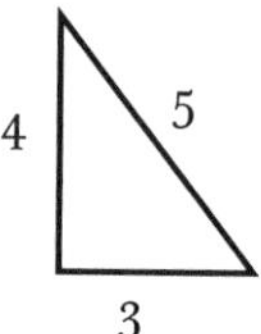

When we multiply each side by 2, we double the length of the sides, but we do not change the basic shape—it is still a right triangle. The two triangles are *similar*, which means that they are scale models of each other.

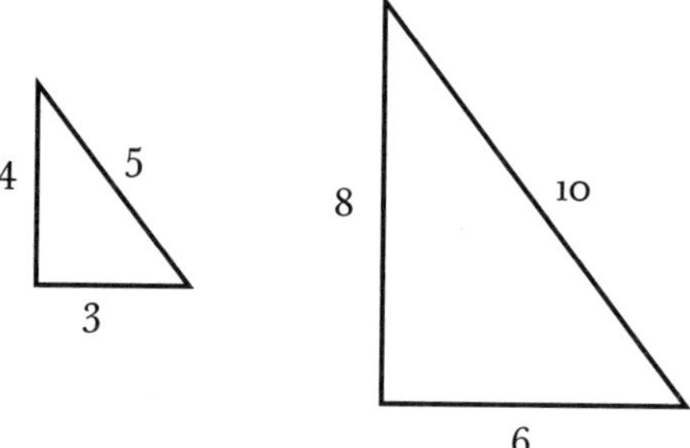

What is important for our purposes is that the new triangle, as a scaled-up model of the original, is still a right triangle. As such, its sides *must* satisfy the Pythagorean relationship. Therefore, it must be that $6^2 + 8^2 = 10^2$ *without even having to check it!* Moreover, there was nothing special about multiplying 3, 4, and 5 by 2. We could have multiplied by any number we wished, and the new triangle would have been a scaled-up model of the original 3-4-5 triangle, meaning that the new triangle would also have been a right triangle. As a right triangle, the sides would have to satisfy the Pythagorean relationship.

Whether we use an algebraic or a geometric argument, the result is that there are an infinite number of Pythagorean triples. Once we find one example, we can generate others simply by multiplying each side by the same whole number. The 3-4-5 triple, however, is known as a *primitive Pythagorean triple*, meaning that there is no common factor among the three numbers. It is one of the "basic" triples from which we can derive others. The triple 6-8-10 and the triple 39-52-65 are not primitive because they are based on the 3-4-5 triangle.

There are other primitive triples. The triple 5-12-13 is almost as famous as 3-4-5. You should check that the sum of 5^2 and 12^2 is indeed 13^2. Like before, we can generate other non-primitive triples from the primitive 5-12-13 one. There are an infinite number of Pythagorean triples, but are there an infinite number of *primitive* ones?

Euclid answers this question for us in Book X of the *Elements*. There, he provides a way to generate not simply an infinite number of primitive Pythagorean triples but *all* primitive Pythagorean triples. The process starts by choosing an n and an m, both positive integers with m being greater than n. Then set the values of a, b, and c accordingly.

$$a = m^2 - n^2$$
$$b = 2 \times m \times n$$
$$c = m^2 + n^2$$

Checking to see that $a^2 + b^2 = c^2$ requires a bit of algebra, so I will leave it to those brave souls who want to complete the proof. Instead, we will simply pick some values to see how the formula works. Let's take $m = 5$ and $n = 3$.

$$a = 5^2 - 3^2 = 25 - 9 = 16$$
$$b = 2 \times 5 \times 3 = 30$$
$$c = 5^2 + 3^2 = 25 + 9 = 34$$

We can now check to see that the 16-30-34 triple satisfies the Pythagorean relationship: $16^2 + 30^2 = 34^2$. With different values of m and n, we can see that Euclid's amazing set of formulas can generate an infinite number of Pythagorean triples. What is not clear is that these formulas generate *primitive* triples. In fact, quite the opposite was true using $m = 5$ and $n = 3$. These values did not produce a primitive triple because 16, 30, and 34 have a common factor of 2. That said, we can obtain a primitive triple by dividing out by the greatest common factor: $16 \div 2 = 8$, $30 \div 2 = 15$, $34 \div 2 = 17$. Therefore, a primitive triple is 8, 15, and 17.

The Geometer provides a way to guarantee that his formulas produce a *primitive* triple. We must choose m and n so that they have no factors in common, and they are not both odd.[143] For example, if $m=12$ and $n=9$, a primitive triple will not appear because 12 and 9 have a factor of 3 in common. Further, if $m=13$ and $n=9$, a primitive triple will not appear, either, because both 13 and 9 are odd. On the other hand, if we choose $m=10$ and $n=9$, these two values have no factors in common and they are not both odd. Therefore, a primitive triple should be produced by Euclid's formulas.

$$a=10^2-9^2=100-81=19$$
$$b=2\times10\times9=180$$
$$c=10^2+9^2=100+81=181$$

I leave it to you to check that 19, 180, and 181 do not have any factors in common and that $19^2+180^2=181^2$. The result, care of Euclid, is that there are not only an infinite number of Pythagorean triples—positive integers a, b, and c that satisfy $a^2+b^2=c^2$—but also an infinite number of *primitive* Pythagorean triples.

At this point, a natural mathematical wondering might arise. What if we change the exponent in the Pythagorean equation? Are there also an infinite number of positive integers that satisfy the cubic version?

$$a^3+b^3=c^3$$

This equation is not a statement about triangles, but it is a natural algebraic extension of the Pythagorean Theorem. Mathematicians wrestled to find positive integer solutions to this equation for quite some time with a surprising lack of results. Not only could they not find an infinite number of solutions, *but they also failed to find even one*. That's right—try as they might, no mathematician was able to find any positive[144] integer values for a, b, and c that satisfy the equation $a^3+b^3=c^3$. The fact that an equation has no positive integer solutions is not itself surprising. After all, the equation $3^a=2^b$, which we discussed in the chapter on irrational numbers, can't have positive integer solutions because the left side is always odd (being a product of a bunch of 3s) and the right side is always

143 I will not present an argument for why this is true, but I encourage you to think through it.

144 If *positive* is removed, then the equation does have solutions. One obvious one is $0^3+0^3=0^3$.

even (being the product of a bunch of 2s). What is surprising about our Pythagorean-like equation is that the lack of integer solutions is not nearly as obvious. For centuries, mathematicians could not find solutions, but neither could they successfully reason that no solutions exist.

In fact, the mathematical plot thickens. They encountered the same problem when the exponents were 4.

$$a^4 + b^4 = c^4$$

This one is perhaps more surprising because it looks a bit like the Pythagorean Theorem itself. It can be written as two squares adding to produce a third square.

$$(a^2)^2 + (b^2)^2 = (c^2)^2$$

Surely integer solutions to $a^2 + b^2 = c^2$ could be used to find solutions to $(a^2)^2 + (b^2)^2 = (c^2)^2$. Alas, try as they might, mathematicians could not find any positive integer solutions to $a^4 + b^4 = c^4$, nor could they reason that no solutions exist.

As if that were not troubling enough, they had the very same problem with similar equations.

$$a^5 + b^5 = c^5$$
$$a^6 + b^6 = c^6$$
$$a^7 + b^7 = c^7$$

In fact, they had the same problem for *any* power greater than 2. The *only* solutions they could find were for the original Pythagorean relationship $a^2 + b^2 = c^2$. When n was greater than 2, they were simply unable to find positive integer solutions to $a^n + b^n = c^n$. On the other hand, they could not prove that a solution was *impossible*.

The problem itself comes from the French mathematician Pierre de Fermat, a parliamentary lawyer by trade. In 1635, Fermat made a curious note in the margins of his copy of *Arithmetica*, a collection of algebra problems and solutions by the ancient Greek writer Diophantus.[145] The problem that interested Fermat was in Book II of the *Arithmetica*. Diophantus asks how a square number can be split into the sum of two other

145 We saw Fermat earlier in our discussion of Fermat primes.

square numbers. That is, given a c^2, how can we split it into a^2+b^2? It is sort of a Pythagorean Theorem in reverse.

In exploring this problem, Fermat wrote in the margin (in Latin), "It is not possible to separate a cube into two cubes, a fourth power into two fourth powers, or in general any power greater than 2 into two like powers."[146] What comes next is outstanding or obnoxious, depending on your mood. Fermat writes that he has a "remarkable proof" (*demonstrationem mirabilem*), but *Hanc marginis exiguitas non caperet*: "The smallness of this margin cannot contain it."

Fermat left many such notes and conjectures in his books. Over the years, mathematicians took to proving those for which he himself did not offer a proof. This one, however, eluded everyone and over time became the last one left. For that reason, it became known as Fermat's Last Theorem. The mathematical translation of Fermat's note is precisely the question at hand.

Fermat's Last Theorem
If n is greater than 2, then there are no positive integer solutions for a, b, and c in the equation $a^n+b^n=c^n$.

The result is shocking. In the case of $n=1$ and $n=2$, not only do solutions exist but there are an infinite number of them.[147] If Fermat is correct, then something happens once the exponent becomes greater than 2 that makes the equation go haywire. We go from an infinite number of solutions to *no solution at all.*

It is not clear that Fermat actually had the proof he claimed to have. Some have speculated that he had an incorrect proof but was not aware of its mathematical flaws. After all, over the centuries many "proofs" have been presented for his marginal theorem only to be found out later to be incorrect. Since Fermat never wrote down his proof, we likely will never know what he had or didn't have. What we do know, though, is that Fermat did eventually write down a proof for the case of $n=4$. In other words, he proved that $a^4+b^4=c^4$ has no positive integer solutions. It is

146 This rough translation of the original Latin is my own. The entire note is "*Cubum autem in duos cubos, aut quadratoquadratum in duos quadratoquadratos et generaliter nullam in infinitum ultra quadratum potestatem in duos eiusdem nominis fas est dividere cuius rei demonstrationem mirabilem sane detexi. Hanc marginis exiguitas non caperet.*"

147 We did not discuss the theorem when $n=1$, but in this case the Fermat equation is simply $a+b=c$, which clearly has an infinite number of positive integer solutions.

easy to minimize the importance of this when comparing it to the full power of Fermat's Last Theorem, but even on its own this proof was a tremendous feat because it demonstrated that at least *one* value of n produces an equation with no solution. And yet, the cases for n greater than 4, together with the case of $n=3$, remained a mystery. The case of $n=5$ was independently proven by the French mathematician Adrien-Marie Legendre in 1830 and the German mathematician Peter Gustav Lejeune Dirichlet in 1828.

It was not until 1993 that a full proof of Fermat's Last Theorem was unveiled by the British mathematician Andrew Wiles, a professor at Princeton at the time. The story of Wiles's proof is itself intriguing, as he worked in relative secrecy on the theorem for almost a decade before unveiling it in a series of three lectures during a mathematics conference at Cambridge. People flooded into the room by the end of the third day as it became clear what was happening. Wiles's proof was huge news, especially in mathematical circles, and was featured on the front page of newspapers across the world. A couple of months later, when the proof began the process of peer review, the energy in the mathematical world was brought to a sudden halt as it was discovered that there were flaws. At this point, Wiles enlisted some help, fixed the proof, and announced success in 1994. This time, his work did survive peer review, and the correct proof was published in 1995.

In his proof, Wiles made use of some very modern and heavy-hitting mathematical structures such as *modular forms* and *elliptical curves*. There is little doubt that, while correct, Wiles's proof is not a proof from *The Book*. In fact, Wiles didn't even prove Fermat's Last Theorem directly. What he really proved was another mathematical conjecture called the Taniyama–Shimura–Weil Conjecture. Mathematicians already knew that if Taniyama–Shimura–Weil was true, so was Fermat's Last Theorem. This connection was conjectured in part by Gerhard Frey and more formally by Jean-Pierre Serre, and then later proven by Ken Ribet. The dramatic story of how this age-old theorem was finally proven raises an interesting question about who deserves credit for the proof. That said, history has spoken, and Andrew Wiles is named as the one who conquered Fermat's Last Theorem.

Whether or not Fermat actually had a correct proof of his theorem, it is clear that he did not have Wiles's proof. Fermat could not have known about the mathematical structures Wiles used. Whether or not Fermat actually had a simpler proof based on the mathematics of his time, it is still a matter of great interest to find such a proof. The theorem itself is easy enough to understand: no positive integer solution exists for the general Pythagorean equation when the exponent is greater than 2. How could such a simply stated theorem not also have a simple proof? Maybe there is a moment of genius yet to come that can provide an insightful proof from *The Book*. Until then, Fermat's Last Theorem remains a theorem about which you should know but for which the proof is not only too big to fit into a margin but also too complicated to fit into this book.

CHAPTER 10
GÖDEL'S INCOMPLETENESS THEOREMS

IN OUR mathematical expedition thus far, we have explored at least a score of different theorems, and the power of mathematics to prove things has become increasingly impressive. We first noticed the ability of a proof to deal with *all* triangles at once even though we have never physically seen any triangles at all. We then witnessed the power of proof to demonstrate that an infinite number of primes exist even though we could only ever write down and know a finite number of them. We even saw how mathematics can demonstrate that certain tasks—trisecting an angle with a straightedge and a compass—are *impossible*, not because we have not yet found a way but because by its very nature the task cannot be accomplished. Perhaps most surprising of all, we described how we can prove that a particular statement such as Euclid's fifth postulate can neither be proven true nor shown to be false and is therefore independent from its four fellow postulates. All of this raises the question about both the power and limits of mathematical reasoning.

Part of why mathematical proof is so powerful is because of the *precision* found at the heart of mathematics itself. Mathematicians insist on being absolutely precise about the definitions with which they work, the axioms with which they start, and the rules of engagement they use in moving from step to step to eventually arrive at the end: a fully articulated true statement with an accompanying assurance of its truth that is unparalleled in other disciplines.[148] In the early twentieth century, a few mathematicians became interested in exploring the formality of mathematics and both the capabilities and limitations of mathematical proof. They were motivated partly by certain paradoxes that had been floating around for a while that, on the surface, seem to be beyond the language and reasoning of mathematics. In 1901, the mathematical philosopher Bertrand Russell published one of the more famous ones.[149]

148 In the *Nicomachean Ethics*, Aristotle comments specifically on the nature of precision and certitude of mathematics as compared with other subjects. Precision, he says, is not to be sought at the same levels in all discussions. It is the "mark of the educated man" to look for precision in each type of argument only so far as the nature of the subject permits. It is foolish, says Aristotle, to accept "probable reasoning" from mathematics, but equally foolish to expect mathematical precision from a rhetorician.

149 Ernst Zermello apparently discovered the same paradox in 1899 but did not publish his result.

Russell's Paradox
Sets are collections of objects. Some of those objects can also be sets themselves. It is even possible for a set to contain itself. Consider, then, the set of all sets that do not contain themselves. Does this set contain itself?

The paradox may be easier to understand if we use an analogy. Let's consider books. Some books refer to themselves, whereas other books do not refer to themselves. For example, consider a book called *A Catalog of All Books About Bats*. The content of this book would contain the titles of any book that is even remotely about bats, such as *Cave Bats: The Dangers of Spelunking*[150] and *The Best Tools for Facing a Vampire's Curveball*. Because our reference book is a book that claims to list *all* books about bats, it too is a book about bats. It therefore makes sense that it would list itself. As such, this book is an example of a text that refers to itself, much like a set that contains itself. Some self-referential books might not be quite as obvious. For example, the book *Another Sort of Mathematics* in the very writing of this sentence is now a text that refers to itself. But most books do not refer to themselves. To the best of my knowledge Evelyn Waugh's *Brideshead Revisited* does not refer to itself directly (apart from the title).

In summary, and at the risk of restating something painfully obvious, *some books refer to themselves* while *other books do not*. Now, as a budding connoisseur of reference books, you might wish to author a text called *A Catalog of All Books That Do Not Refer to Themselves*. Our bat reference book, *A Catalog of All Books About Bats*, would *not* go into this new project because it *does* refer to itself, nor would *Another Sort of Mathematics*. *Brideshead Revisited*, however, would be among those listed in your newly published work.

Consider now whether you should you list the book *A Catalog of All Books That Do Not Refer to Themselves* in the book *A Catalog of All Books That Do Not Refer to Themselves*. The paradox is this: if you decide to include it, then it is now a book that refers to itself and therefore should *not* be included, and yet if you decide not to include it, then by definition it is a book that does not refer to itself and *should* be included. It is a lose-lose situation, and our project of writing *A Catalog of All Books That Do Not Refer to Themselves* is doomed to either inconsistency or incompleteness

150 If the Latin pun occurred to you without reading this footnote, my sincerest compliments.

from the start.[151] This is like trying to construct a set of all sets that do not contain themselves.

In 1910, Alfred North Whitehead and Bertrand Russell published the *Principia Mathematica* in part because of their concern about paradoxes such as this, but primarily to formalize mathematics by giving it a firm, irrefutable foundation. If such a foundation could be provided, perhaps seemingly contrived statements such as Russell's paradox would be exposed as mathematical nonsense, or perhaps a complete formalization of mathematics would be able to deal with these statements through clear definitions of things like "set" or "collection." Their project consisted of reducing the number of assumptions (axioms) for mathematics and converting all mathematical statements to pure symbols in order to eliminate the ambiguity of natural language.

Limiting the axioms is easy enough to understand, but what do we mean by "converting mathematical statements to pure symbols"? For example, rather than stating, "Addition is commutative," or even "For any number a and b, the result of $a+b$ is equal to the result of $b+a$," we would instead write something like "$\forall a{:}\forall b{:}(a+b)=(b+a)$"[152] Using this type of formalization, we could reduce mathematical statements to a relatively small number of symbols. Converting mathematics to symbols is not foreign. The very study of algebra is an exercise in symbolic representation and manipulation. The task of Whitehead and Russell, however, was to eliminate words altogether and develop something of an "algebra of mathematical language." In this new formalized system, steps in a proof could become mere exercises in the manipulation of symbols. For example, instead of thinking about the *meaning* of "$\forall a{:}\forall b{:}(a+b)=(b+a)$" as "the order of addition can be switched," we can see it as a simple instruction about how to change symbols in a mathematical sentence.

> If a proof has "[a first something] + [a second something]" anywhere in a previous step, we can replace that with "[a second something] + [a first something]."

151 There is another famous example related to this paradox. In a particular small town, the barber cuts the hair of everyone who does not cut his own hair. Who cuts the barber's hair?

152 These are not necessarily the symbols and organizing principles used in *Principia Mathematica*.

For example, if a previous step in a proof is "$\exists b:(b+1)=2$" our next step could be switching the 1 with the b: "$\exists b:(1+b)=2$." More importantly, we would not have to worry about the *meaning* of either of these statements when performing the symbol switch. This may seem like mincing words. After all, what does it matter whether we think about the meaning of commutativity or simply about modifying symbols?[153] It actually matters a great deal. The very goal of mathematicians such as Whitehead and Russell was to see whether mathematical proof could be "algorithmicized." While the modern computer had not yet been invented, the study of *algorithm* was of great interest to mathematicians, logicians, and computer scientists.[154] If the starting axioms and definitions are precise combinations of a small number of symbols, and if these axioms also govern the rules for how the symbols can be manipulated from one step to the next, can we simply apply every combination of steps starting with the axioms and eventually generate true theorems, and maybe even *all* true theorems? Moreover, the very generation of the theorem would also provide the proof! We wouldn't have to *think* about how to prove a theorem; we would only have to apply steps blindly according to the rules. Of course, the modern implication for this extreme formalization would be the existence of a computer program that could produce both mathematical theorems and their proofs. This question is more relevant than ever with the development of artificial intelligence and large language models.

Think of it this way. Suppose that we have just six axioms assembled using a relatively small number of symbols: x, y, z, *, +, =, 0, 1, (,), and $\Rightarrow$.[155]

153 Please excuse the pedagogical excursion in this footnote. The removal of meaning from symbolic manipulation is precisely the problem with how algebra, and much of mathematics, is taught to students. They misapply properties, such as replacing $(a+b)^2$ with a^2+b^2, because they have been trained to memorize rules of symbolic manipulation without pausing to think about the meaning of these symbols. Algebra teachers can be far more effective if they consistently use language about numbers and their properties instead of language about symbols. Even the *raison d'être* for doing algebra should be the need to represent "all numbers" or an "unknown number."

154 The study of computation by "computer scientists" existed long before the advent of the modern computer.

155 There is a lot of work to do when deciding which strings are "well defined" in our system. This is especially true for adding and removing parentheses. I have tried to keep the presentation here as simple as possible. Notice in the proof that follows that I left "(1 * 0)" with the parentheses. This is because a rule for removing parentheses is harder to codify than the rule for adding them. Even the way I have stated the axiom for adding parentheses is an oversimplification.

#	Symbolic Axiom	Mathematical Meaning	Symbolic Manipulation
1	$x=x$	Any number equals itself.	We can start with any string, followed by "=", followed by the same string.
2	$x \Rightarrow (x)$	Adding parentheses around an expression doesn't change the expression.	"(" can be placed at the front of any string so long as ")" is also placed on the end.
3	$x \Rightarrow 1*x$	Multiplying a number by 1 doesn't change the number.	"1*" can be placed before any other string.
4	$x \Rightarrow 0+x$	Adding 0 to a number doesn't change the number.	"0+" can be placed before any other string.
5	$x*(y+z) \Rightarrow (x*y)+(x*z)$	Multiplication distributes over addition.	If three strings are in the form of $x*(y+z)$, the entire thing can be replaced by $(x*y)+(x*z)$.
6	$x+z=y+z \Rightarrow x=y$	If two wholes are equal and share the same part, then the remaining parts are also equal.	If a string has an equal sign and "+ [some string]" that occurs at the end of both sides, it can be removed from both sides.

Let's see how we can generate a "theorem" in this system by starting with the second axiom.

Statement	Rationale
0=0	We can start with this per Axiom 1.
(0)=(0)	Place parentheses (twice) per Axiom 2.
1*(0)=(1*0)	Place "1*" (in two different locations) per Axiom 3.
1*(0+0)=0+(1*0)	Place "0+" (in two different locations) per Axiom 4.
(1*0)+(1*0)=0+(1*0)	Replace the left-hand string per Axiom 5.
(1*0)=0	Remove "+(1*0)" from both sides per Axiom 6.

We have a relatively elaborate proof that 1 times 0 equals 0, but what is important is that our "reason" for each of the steps is described in terms of what happens to the symbols and not in a way that is directly associated with mathematical reality. "Place '0+' onto a string" is something very different than "Adding zero to a number doesn't change the number." Once we realize that the steps can be dissociated from their meaning and described only in terms of symbols, we could write an algorithm—what we now associate with a computer program—that starts with any axiom and methodically applies the other axioms as a mere manipulation of symbols at each step to generate true theorems *together with their proofs*!

As I said, though, there was something deeper at work in Whitehead and Russell's project than just algorithmicizing mathematical proof. It was really an attempt to secure the very foundations of mathematics and mathematical reasoning. Around that time, the German mathematician David Hilbert was concerned about similar things. In what is known as "Hilbert's Program," he asked for the same formalization for all of mathematics as Whitehead and Russell, but he specified that the formalization should be able to show that mathematics is both *complete* and *consistent*. *Complete* means that any true statement in mathematics has a proof, even though we may not have discovered it yet. *Consistent* means that no contradiction can be arrived at, i.e., no statement can be proven to be both true and false. Hilbert challenged the mathematics community in 1900 during a monumental talk at the International Congress of Mathematicians in Paris. With the advent of a new century, he proposed a list of ten problems that he thought should occupy mathematicians for the next one hundred years.[156] One of these problems, the second on the list, asked mathematicians to prove the *consistency* of the axioms of arithmetic. The challenge of *completeness* arrived in 1928, when Hilbert specifically issued his *Entscheidungsproblem*—roughly translated as "the Decision Problem"—to find an algorithm that determines whether any mathematical statement is true or false. If such an algorithm exists, the axioms of mathematics could be considered "complete."

In 1931, a twenty-five-year-old Austrian logician named Kurt Gödel answered Hilbert's challenge, but in a way that shook mathematics to its very core. Hilbert asked for a formalized system of mathematics that was both consistent and complete, meaning that it was capable of providing a proof for all true statements without also providing "proofs" of false statements. Gödel answered by proving that this task was impossible. The result is known as Gödel's First Incompleteness Theorem.

> *Gödel's First Incompleteness Theorem*
> No formalized system of arithmetic can be both complete and consistent. In other words, so long as the system is not self-contradictory, there will always be true mathematical statements for which a proof is impossible.

The First Incompleteness Theorem is probably one of the most misunderstood theorems in all of mathematics. In order to fully appreciate how shocking this theorem is, it is worth spending some time on what it is *not*. At the extremes, it is often stated as "There are some truths beyond the reach of mathematical proof, for example the existence of God." True though this statement may be, it is not what Gödel's theorem says. Sometimes it is also mischaracterized as "There will always be open problems in mathematics." Again, true though this may be, open problems are

156 Hilbert actually had twenty-three problems in mind, the complete list of which was published a few years later, but only ten were discussed at the Paris lecture.

open because solutions have not yet been found, not necessarily because solutions *will never be found*. Finally, Gödel's theorem is sometimes related to the kind of *independence* of a particular statement we saw with Euclid's Parallel Postulate. In the chapter on the independence of this postulate, we showed that the first four postulates cannot be used to prove the fifth or its opposite. In some sense this does mean that a proof does not exist for the Parallel Postulate. Gödel, however, is saying something significantly different. First, whether or not the Parallel Postulate is "true" depends on the underlying definitions of the particular geometry. It relies on terms that need definitions. Is a *plane* "a surface which lies evenly with the straight lines on itself," which is Euclid's definition, or is it the surface of a sphere? Being independent from the other four postulates, it might be true or false based on how the basic objects of geometry are defined. The unprovable statements Gödel described are categorically different from these examples. Their truth does not depend on definitions. They are assuredly *true* statements that are yet *unprovable*. Second, with the Parallel Postulate, once we know that it is independent of the other four, *we simply add it to the list*. This is in fact what Euclid did—though he may not have been fully aware of its independence. With Euclid's full set of five postulates, we actually *do* have a geometric system that is both consistent and complete, a fact proven by Alfred Tarski, a Polish American logician in the twentieth century.

The situation for arithmetic is far more dire. Gödel says that *no matter what system of axioms we adopt for arithmetic, there will always be true statements for which a proof is impossible*. We can continue adding those unprovable "Gödel statements" to our system, but even that expanded system will simply generate more unprovable Gödel statements. It is in the very nature of consistent arithmetic to be *incomplete*. Moreover, as we shall soon see, the Gödel statements from arithmetic that are not provable can be known by way of logical reasoning to be true. In other words, we end up with statements *known to be true* but for which formal mathematics is unable to provide a proof. This situation is decidedly more interesting than the independence of the Parallel Postulate.

The brilliant thing about Gödel's work is that he provided a mathematical proof that demonstrates the limits of mathematical proof itself. How did he accomplish this feat? While the proof is very difficult, we can sketch the basic ideas. Gödel begins with a self-referential sentence not altogether different than Russell's Paradox.

This statement cannot be proven using the axioms of arithmetic.

Assuming arithmetic is consistent—meaning that proofs cannot be provided for false things—this statement *must be true*. If it were false, then by the very statement itself, it would have a proof. But we would then have a valid proof from the axioms of arithmetic showing that a false statement is true, which violates consistency. Therefore, the statement cannot possibly be false.[157] Being *true*, however, by its very content means that there can be no proof for it. That is, after all, what the statement means! Again, this situation is different from the case of the Parallel Postulate; we have just argued using logical reasoning that the statement must be *true*. We can logically demonstrate its truth apart from any formal mathematical proof. Moreover, by the content of the statement itself, it means there can be no mathematical proof for it, despite its *being true*. At first, this may not strike us as all that profound. At the very least the statement seems more like a linguistic trick. If this statement is the only thing that is true but not provable, then why would it shake mathematics to its core? The impossibility of *A Catalog of All Books That Do Not Refer to Themselves* does not impact the legitimacy of all reference books, let alone all books. Moreover, besides its potential singularity, *this statement is not really a mathematical statement!*

To answer these objections, the first moment of Gödel's genius came when he found a way to transform his self-referential statement into the kind of formal mathematical symbols envisioned by Whitehead, Russell, and Hilbert. In other words, he was able to write "This statement cannot be proven using the axioms of arithmetic" *as a mathematical statement*. We are now one step closer to a troubling incompleteness. With the translation, Gödel now had a statement involving only the symbols of mathematics that is true but for which there could never be a proof. Its content, however, is still a bit bizarre and certainly would not worry mathematicians. What Gödel did next was, if such a thing is possible, beyond genius. He took the system of symbols used to code the axioms

157 This is why Gödel's theorem says that an axiomatic system cannot be both consistent and complete. The only way to rescue the self-referential statement from leading to incompleteness is to assume that the axioms are inconsistent. If they are consistent, then the proof continues.

and definitions of formal mathematics and started assigning numbers to them.[158] Therefore, a statement such as "0+0=0" can now be described as a number, for example 79,059,316.

Not only could he translate every mathematical statement into a number, but he could also describe valid proof steps in terms of basic numerical operations. We saw earlier, in our overly elaborate proof that 1*0=0, how to think about the steps in a proof in terms of symbolic manipulation rather than their actual mathematical meaning. For example, the axiom "$x \Rightarrow 1 * x$" might *mean* "Every number multiplied by 1 is itself," but it can be thought of as "The string '1 *' can be placed before any other string." By translating the symbols into numbers in which, for example, 5 represents "1" and 7 represents "*", the axiom might read something like "The digits 57 can be placed in front of any other string of digits."[159]

The thing that makes the manipulation of numbers different from the manipulation of words is that the former can always be described in terms of mathematical operations. For example, if the previous step, as a Gödel number, is 3,477,213, then applying the axiom produces as a next step 573,477,213. As a movement from one number to another number, this "axiom" can be interpreted mathematically as "add 570,000,000."[160] In other words, the axioms of mathematics can be described not only as numbers themselves but also as *operations on numbers*. We can then take a mathematical conjecture, code it as a Gödel number, and ask, "Can we arrive at this number by applying a series of operations represented by the axioms?" Because the number and the symbols are convertible back and forth, asking this is the same as asking, "Is there a proof for our mathematical conjecture?"

We now return specifically to Gödel's self-referential statement: "This statement cannot be proven using the axioms of arithmetic." He found a way to code this statement using formal mathematical symbols and then to turn it into *a statement about numbers*. Because the original statement is *true*—which we argued logically—the converted statement into

158 He actually used prime numbers for each of the symbols, for reasons that have everything to do with the Fundamental Theorem of Arithmetic. Using unique factorization, he could guarantee that each statement had a unique number.

159 The number 57 is not actually Gödel's number for "1 *", but it serves as an appropriate example of what Gödel was doing when assigning numbers to symbols.

160 Our actual axiom involved placing 57 anywhere in the string, no matter how large the string is. The corresponding mathematical operation is a bit harder to describe, but it can be done.

mathematical symbols is true, which means the converted statement about *numbers* is also true. But because the original statement is *unprovable*, the converted statement into mathematical symbols is unprovable, which means the converted statement about numbers is also unprovable. We now have a statement purely about numbers that we *know to be true* but *unprovable*.[161]

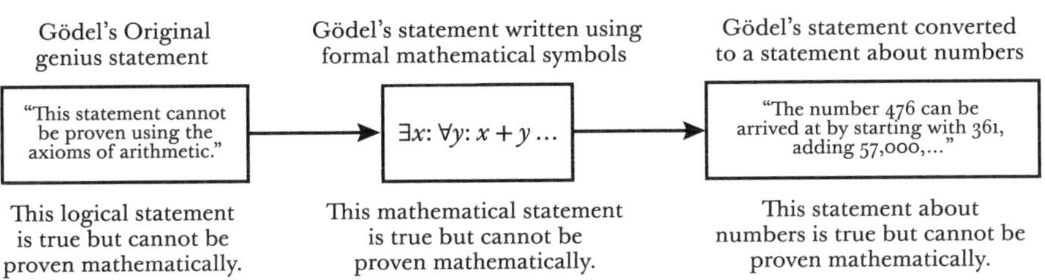

Moreover, there is not just *one* of these statements. If we change the way we assign numbers to each of the symbols, we get a different *true* but *unprovable* statement. With an infinite number of ways of assigning numbers to symbols, we will have an infinite number of true mathematical theorems for which a proof is impossible.

It is worth reflecting on the power of what Gödel did: he made it so that math can talk about itself mathematically. It is also worth, for the sake of absolute clarity, restating the incompleteness theorem and making sure to emphasize the fact that it is about actual *mathematical* statements, not just a one-off, self-referential statement.

> *Gödel's First Incompleteness Theorem*
> No formalized system of arithmetic can be both complete and consistent. In other words, so long as the system is not self-contradictory, there will always be an infinite number of *true mathematical statements* for which a proof is impossible.

There seem to be stages that mathematicians go through when encountering Gödel's theorem. At first, they imagine the very foundations of the discipline crumbling. Maybe a mathematician has been working on a

161 Of course, the symbols in this graphic as well as the particular formulation of the "statement about numbers" is entirely fabricated. Gödel's actual translation into mathematical symbols and then into a statement about numbers is extremely complicated and difficult even to write down.

problem for a very long time and has been unable to find a proof. Before Gödel, in the face of a difficult theorem, the mathematician practices fortitude: "If at first you don't succeed, try again!" After Gödel, that same mathematician might begin to worry: "Maybe it isn't the case that I have not tried hard enough. *Maybe this problem is a Gödel statement.* Maybe it is *true* but not *provable*."

Someone in the second stage, with a bit of nervous laughter, might say, "Well, even though Gödel showed that unprovable true mathematical statements must exist, it seems like they are convoluted statements about very large numbers. It is highly unlikely to affect the normal questions mathematicians ask." The problem with this train of thought, and the source of the nervous laughter, is that mathematicians now know of at least a couple of "natural" Gödel sentences. One of them is known as the Strengthened Finite Ramsey Theorem and concerns being able to categorize (or "color") certain subsets of the natural numbers.[162] A second one is known as Goodstein's Theorem, which asks whether certain sequences of numbers eventually end at 0. Both theorems are perfectly reasonable statements about numbers and are certainly the type of thing a mathematician might try to prove. Therefore, this second stage of comfort is untenable. The third stage is a return to the first. The mathematician becomes quite concerned about any problem that has thus far eluded proof. No one wants to spend a lifetime working on a problem that ends up being unprovable, not because it is false, but precisely because it is *true* yet evades proof altogether. The real issue is that the mathematician might not know whether a hard theorem is an example of a Gödel statement. In fact, for a while, some people started to think that Fermat's Last Theorem might be an example, and therefore unprovable. We now know, however, that this is not the case: the theorem has been proven by Andrew Wiles.

At this point, in the final stage of reckoning with Gödel's theorem, we come to the crux of the matter. Mathematicians have shown that the Strengthened Finite Ramsey Theorem and Goodstein's Theorem are Gödel statements, but remember what a Gödel statement is. It is a *true* statement for which formal mathematics will always be unable to provide

162 The Strengthened Finite Ramsey Theorem is a generalized version of the Friends and Strangers problem that will show up in the next part of our journey: "Unsolved Problems Worth Knowing." The "colors" here would refer the categories of relationships. In the Friends and Strangers problem, there are only two "colors": one to represent being friends and one to represent being strangers.

a proof. This means that *precisely as Gödel statements*, the Strengthened Finite Ramsey Theorem and Goodstein's Theorem are true! Oddly enough, *in proving that these two theorems are Gödel statements, we have provided a proof that they are in fact true!* It's as if proving that these theorems are unprovable is the very way in which we proved them to be true.

If your head is spinning at this point, welcome to the club of those who have tried for the first time to wrestle with Gödel's Incompleteness Theorem. It says that there are true statements about mathematics that will never be able to be proven, but as soon as we specifically identify one of those statements, then by its mere identification we know it to be true—and this seems to suggest that we have proven the very statement which was supposed to be unprovable. The way out of this conundrum is to remember that Gödel's theorem is about *extremely formalized mathematics*, the kind that can be performed by a computer, and not about mathematics generally. *Formal* mathematics, as envisioned by Whitehead, Russell, and Hilbert, will always encounter true mathematical statements that escape *formal* proof. They do not, however, escape proof altogether. It is just that we have to step outside the formal system to be able to provide the proof. Remember, we provided a *logical* proof that Gödel's original self-referential statement was true, even though algorithmic mathematics could never do the same.

The idea of stepping outside the formal system is more obvious in Gödel's *Second* Incompleteness Theorem.

> *Gödel's Second Incompleteness Theorem*
> No formalized system of arithmetic can ever
> demonstrate that its axioms are consistent,
> that is, that they do not lead to contradiction.

It isn't that we cannot *know* that arithmetic is consistent, and it isn't even that we cannot argue, or even prove, that it is consistent. It is just that extremely *formalized* arithmetic cannot prove its own consistency. In order to do demonstrate consistency, we need to step outside the formalized system.

Gödel did not shatter the foundation of mathematics itself, but he did shatter the hopes of those seeking to turn mathematics and mathematical proof into a mere application of algorithms. Gödel definitively answered Whitehead, Russell, and Hilbert by saying that mathematical proof is more than a formalized algorithmic system. It must involve something other than a mere step-by-step process; it must involve moments of logical *insight*. In this way, there are also profound anthropological ramifications for Gödel's work, something he himself recognized. The result of the First Incompleteness Theorem is that we can know a statement to be true and even have a logical proof that it is true, but this proof could never be the result of a series of automated steps taken by a computer. It is not too much to say that one implication of Gödel's work is that the human mind is not purely algorithmic and is therefore not a computer.[163] The British mathematician Roger Penrose, in thinking about mathematical truth and the significance of Gödel's work, asks,

> *How do we form our judgements as to what is true and untrue about the world? Are we simply following some* algorithm*? ... Or might there be some other, possibly non-algorithmic route—perhaps intuition, instinct, or insight—to the divining of truth?*[164]

Penrose sees an answer in Gödel's theorem:

> *Mathematical truth is not something that we ascertain merely by use of an algorithm. I believe, also, that our* consciousness *is a crucial ingredient in our comprehension of mathematical truth. We must "see" the truth of a mathematical argument to be convinced of its validity. This "seeing" is the very essence of consciousness. It must be present* whenever *we directly perceive the mathematical truth. When we convince ourselves of the validity of Gödel's theorem we not only "see" it, but by so doing we reveal the very non-algorithmic nature of the "seeing" process itself.*[165]

163 Alan Turing was working on the Hilbert problem around the same time, but from the perspective of the Theory of Computation. In his famous "Halting Problem," he showed that no algorithm (or computer program) could ever be developed that determines whether other algorithms eventually stop or go on forever. Like the Gödel sentence, human reason can be used to know that the particular "foil" to a supposed halting program actually does halt, even though the algorithm itself cannot determine this, again showing that the human mind must have some non-algorithmic component to it. To the best of my knowledge, Turing never admitted this anthropological consequence of his theorem, whereas Gödel did.

164 Penrose, Roger. *The Emperor's New Mind: Concerning Computers, Minds, and the Laws of Physics*. Oxford University Press (New York, NY), 1989, page 99.

165 Penrose, page 418.

This type of "seeing" is precisely what separates human reason from a computer's algorithmic process. It is also what ensures that "A.I." will always be "A." and never "I." The kind of proof that results from such insight rather than an algorithm seems delightfully close to what Erdős meant by a proof from *The Book*.

PART IV

TEN UNSOLVED PROBLEMS WORTH KNOWING

CHAPTER 1
THE PERFECT BRICK

THERE IS a challenging problem commonly found in mainstream mathematics textbooks goes something like the following.

Find the length of the line that passes through the box from one corner to the opposite corner.

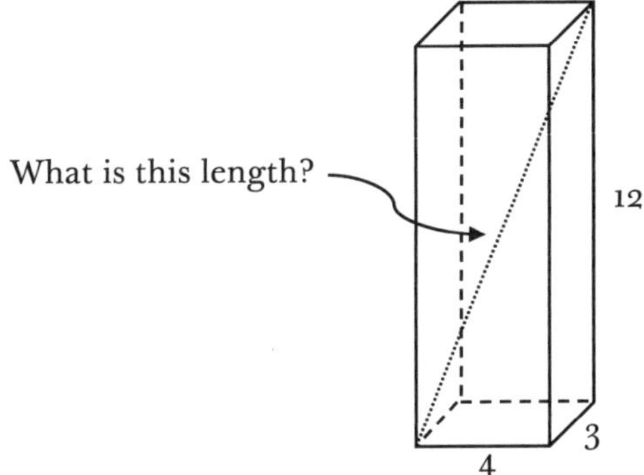

In fact, George Pólya, a Hungarian mathematician and pedagogue, presents this problem as a main example in his book *How to Solve It*,[166] which is both a manifesto and a manual on teaching problem solving to students of mathematics. Pólya phrases the problem in general terms.

Find the diagonal of a rectangular parallelepiped of which the length, the width, and the height are known.

Pólya uses the problem both to illustrate the importance of giving students novel problems to solve—as opposed to repetitive exercises—and to demonstrate how Socratic questioning can unfold in an attempt to lead the student while still leaving him to do the heavy lifting and creativity necessary for learning. One such Socratic question Pólya suggests is "Do you know a related problem?" For the problem about the diagonal of a box, where have we seen this sort of thing before? Very early on we saw a related problem in the context of the Pythagorean Theorem. In this new context, though, our box does not have any triangles. In a Tantonian move, when we don't have something, we make it happen.

166 Pólya, George. *How to Solve It: A New Aspect of Mathematical Method*. Princeton, NJ: Princeton University Press, 1957. In *How to Solve It*, Pólya also includes the problem of the traveling bear that I used to illustrate geometry on the surface of a sphere.

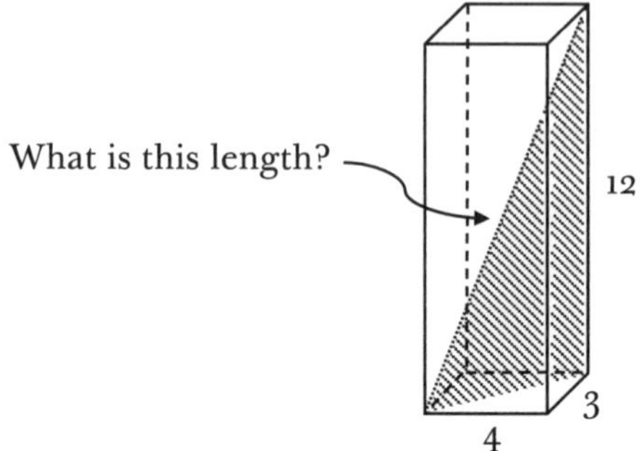

Because the box is formed with all right angles, the shaded triangle is a right triangle, and therefore a candidate for the Pythagorean Theorem. We could find the length of its diagonal—the hypotenuse—if only we knew the length of its two legs. Unfortunately, we only have the length of the tall leg, 12.

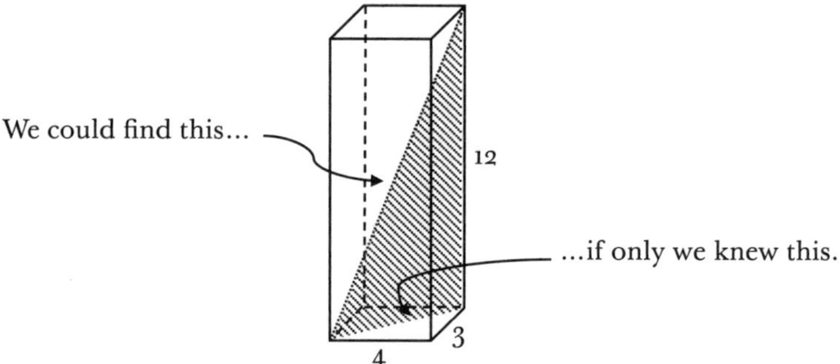

If we had an actual three-dimensional model, it would be clear that there is at least one more right triangle in the picture.

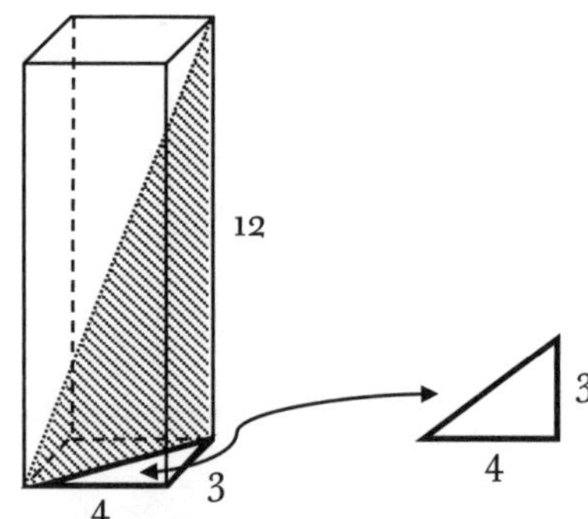

The missing hypotenuse of this smaller triangle, when squared, must equal 3^2+4^2, or 25. Therefore, the hypotenuse must be 5. (You might remember from the chapter on Fermat's Last Theorem that 3-4-5 is the most basic of Pythagorean triples.)

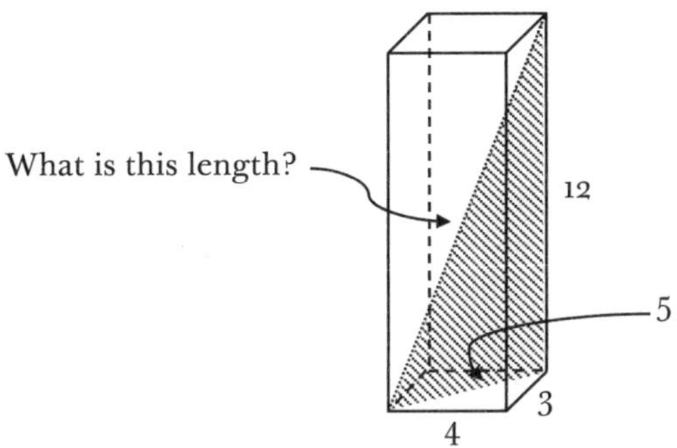

We are now ready to deal with the larger right triangle, which has legs equal to 5 and 12. I will leave it to you to work out, but these legs form the first two parts of another Pythagorean triple, with the hypotenuse being equal to 13. We have thus solved our problem by using the Pythagorean Theorem *twice*. The distance from one corner of the box to the opposite corner is 13.

Pólya's question was not about a specific box; it was more general than that. He does not provide lengths and still asks how we might determine the lengths of the main diagonal. But using our specific example, we now know that we can apply the Pythagorean Theorem twice to find the desired length. It may, however, be worth our time to write down the more general relationship. While commonly expressed as an equation, we made a big deal earlier about the fact that the Pythagorean Theorem is really about areas of squares. It says that the square on the hypotenuse is equal in area to the sum of the squares on the legs. If we draw our two right triangles and label their regions, it might look like this. (I will keep the same shading and bold lines so that you can more easily compare them with our original problem.)

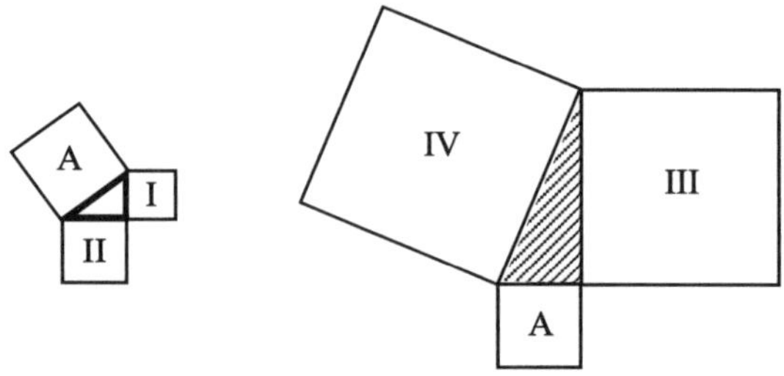

The regions marked I, II, and III are squares on the original sides of the box. I and II are built on the length and width of the base, and III is built on the height. The region marked IV is the square built on the main diagonal. According to the Pythagorean Theorem, we know that A is equal to I + II. According to the second application of the Pythagorean Theorem, IV is equal to A + III. We can glue both equations together.

IV = A + III

‖

I + II

Therefore, IV = I + II + III. If we would like to write this equation as one involving the squares on the original length, width, and height of the box, using d for the main diagonal, we would have $d^2 = l^2 + w^2 + h^2$. If we use a, b, and c for the box dimensions, we arrive at a theorem beautifully similar to the Pythagorean Theorem.

> *The 3-D Pythagorean Theorem*
> If a, b, and c, are the three dimensions of a box and d is the diagonal from one corner of the box to its opposite corner, then $d^2 = a^2 + b^2 + c^2$.

We call this theorem the 3-D Pythagorean Theorem because it really is a generalization of its two-dimensional counterpart. Although the original theorem is about squares built on the sides of a triangle, it can also be seen as a theorem about the diagonal distance between two points on a plane when you know horizontal and vertical distances. Likewise, our 3-D Pythagorean Theorem is about squares on the edges of a box, but it is also about the distance between two points in space when you know the three component distances.[167]

Notice that our opening example had three box measurements that were all whole numbers and a diagonal that was also a whole number. Such a situation is called a Pythagorean quadruple, akin to the Pythagorean triples that we met in the chapter on Fermat's Last Theorem.

> DEFINITION: A *Pythagorean quadruple* is a group of four whole numbers a, b, c, and d such that $d^2 = a^2 + b^2 + c^2$.

The group (3, 4, 12, 13) is one example of a Pythagorean quadruple. Other examples are (1, 2, 2, 3), (2, 3, 6, 7), and (1, 4, 8, 9). In fact, these are all *primitive* Pythagorean quadruples, e.g., where a, b, c, and d have no common factor except for 1. As in the case of Pythagorean triples, there are an infinite number of Pythagorean quadruples and even an infinite number of primitive Pythagorean quadruples.

Our box has something else interesting about it. Not only are the lengths of the sides and the main diagonal whole numbers, but the diagonal of the bottom (and top) of the box is also a whole number: 5. What about the other *face diagonals*?

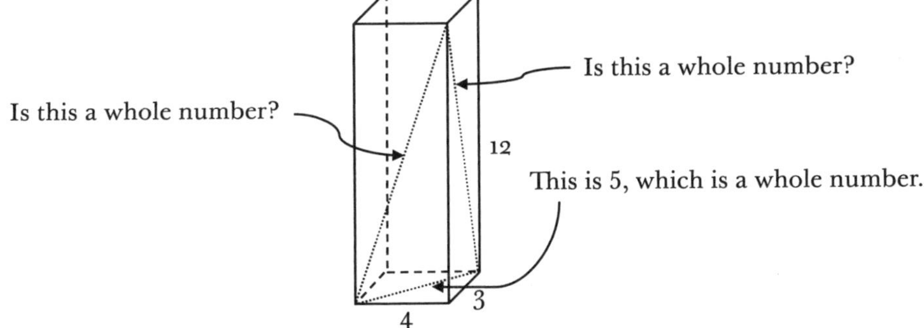

167 The formula actually generalizes to dimensions higher than three. This book is not the place, however tempting it may be, to present higher-dimensional geometry and what that means.

The two unknown face diagonals are, not surprisingly, each a hypotenuse of a right triangle.

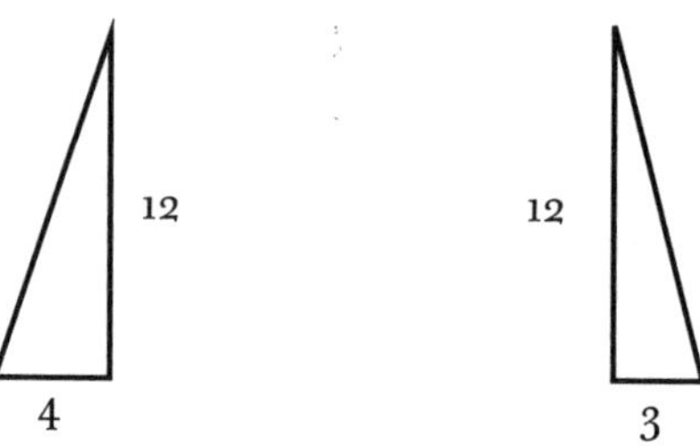

I will let you work it out, but the length of the diagonal of the front face is $\sqrt{160}$, and the length of the diagonal of the side face is $\sqrt{153}$, neither of which is a whole number.

There *are* examples of boxes that have whole-number sides and all whole-number diagonals on each of the three faces. For example, a box with a length, width, and height of 240, 252, and 275 has face diagonals of 348, 365, and 373, all whole numbers. You should check these calculations.

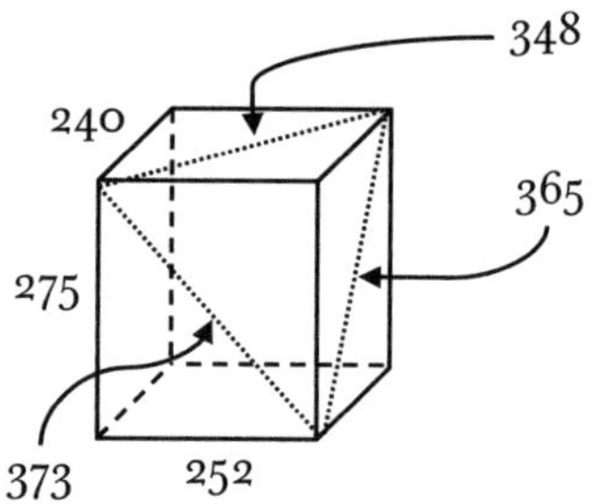

These types of boxes are known as *Euler bricks*.

> DEFINITION: A *Euler brick* is a three-dimensional box that has all whole-number sides and all whole-number face diagonals.

When we multiply each side by a scale factor, the diagonal is multiplied by the same scale factor. Therefore, once we have one Euler brick, we have an infinite number of them. Euler himself was able to prove that there are an infinite number of *primitive* Euler bricks: whole-number sides, whole-number face diagonals, and the only common factor for the sides being 1.

What about the main diagonal of our (240, 252, 275) Euler brick? Is it a whole number?

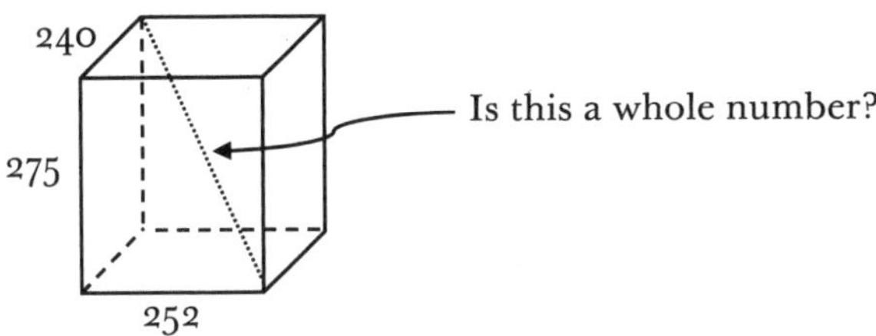

We now know the 3-D Pythagorean Theorem. If d is the length of the main diagonal, then $d^2 = 240^2 + 275^2 + 252^2$, which is 196,729. Unfortunately, this number is not a perfect square, so d is not a whole number. (It is approximately 443.54.)

Our first box had all whole-number sides and a whole-number main diagonal, but not all whole-number face diagonals. Our second example had all whole-number sides and all whole-number face diagonals, but not a whole-number main diagonal. A box that has whole numbers for all of these lengths is called a *perfect Euler brick*.

> DEFINITION: A *perfect Euler brick* is a rectangular box that has all whole-number sides, all whole-number face diagonals, and a whole-number main diagonal.

We know of boxes that are very close to perfection. Both of our examples have six out of seven lengths that are whole numbers. A third example is the $(520, 576, \sqrt{618{,}849})$ brick, which also manages to have six out of seven measurements that are whole numbers, including all face diagonals and the main diagonal. Unfortunately, it has a side that is irrational because 618,849 is not a perfect square.

Does a perfect Euler brick exist? Astonishingly, we do not know. Mathematicians have not found one but also have been unable to prove that one cannot exist.

> UNSOLVED PROBLEM: Does a perfect Euler brick exist?

Maybe Pólya's advice could be helpful in someday solving the problem.

Do we know a related problem: the Pythagorean Theorem, Fermat's Last Theorem, or something else from mathematics? Until then, it remains an open question.

CHAPTER 2
THE NAPKIN FOLDING PROBLEM

IN PART III of this book, we saw that one of the limitations of Euclidean geometry was the types of shapes that can be constructed using a straightedge and a compass. It seemed counterintuitive that *trisecting* an angle was so difficult, and eventually proved impossible, especially when *bisecting* an angle is easy enough that it is a standard part of most high school geometry courses. The impossibility of trisection has in part to do with the two chosen tools, a choice that is codified in Euclid's postulates. What if the tools change? What if the geometric universe is not so much about lines and circles, but instead about *reflections*?[168] For example, what if you wanted to reflect the following shape across the line? What "tool" might you use to accomplish the task?

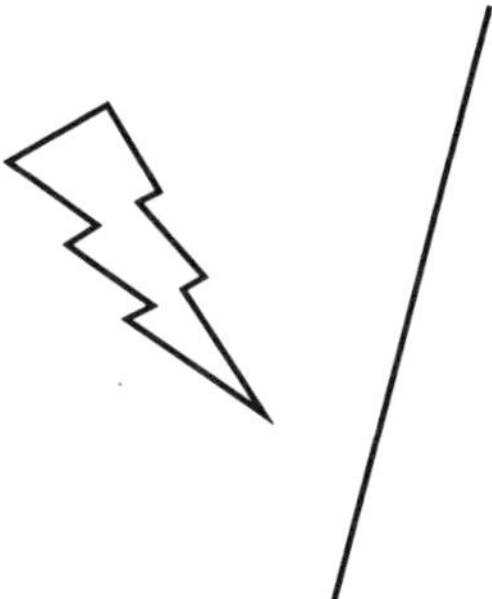

Children learn about reflections by folding the paper and tracing the result. Folding along a line provides the "tool" to perform a reflection—no straightedge or compass required.[169] It turns out that with paper folding—rather than a compass and a straightedge—it *is* possible to trisect an angle, one of the tasks that eluded constructions using Euclid's tools.

168 This is not unreasonable. While I did not include a chapter on rigid motions in this book, they are worthy of at least a footnote. In order to codify the concept of *congruence*, Euclid really needs the idea of a *rigid motion*, which is a mapping of one shape to another so that the distance between any two points is preserved. In other words, if points *A* and *B* are on a shape and two units apart, when we move the shape, *A* and *B* are still two units apart. There has been no stretching, no shrinking, and no tearing. Sliding (or *translation*) is a rigid motion. It is moving a shape from one location on a table to another in a straight line. *Rotation* is also a rigid motion. The unique rigid motion, however, is *reflection*. When we reflect an image, the result is backwards, but the general shape is preserved. So, the preimage and the image are congruent. It is a beautiful result of geometry that *any rigid motion that you can possibly think of*—be it a rotation, a translation, a combination of the two, or some other rigid motion we have not yet named—*can be thought of as the result of three or fewer consecutive reflections*. In other words, take a shape and mark its outline on a table. Throw the shape randomly across the table, flipping it to allow for the result to possibly be backwards, and mark the result on the table. You will always be able to get from your first shape to its randomly thrown image in 0, 1, 2, or 3 reflections. No sliding or rotating will be needed. Therefore, *reflections* are the basic building blocks of rigid motions. They are the "prime numbers" of congruence itself. The proof of this result is a top candidate for a second volume!

169 A mirror would give the same result, but we would have to carefully define a reflection in that context. For my purposes in this chapter, using paper folding will be much more convenient.

Consider an arbitrary angle on a square piece of paper.

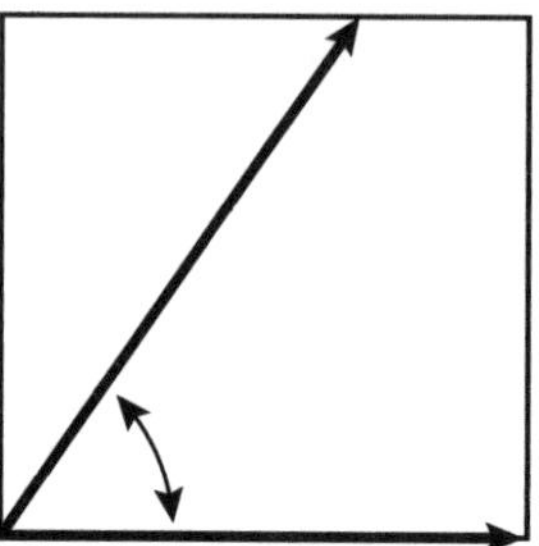

First, fold the paper so that the top meets the bottom, making a crease that is halfway up the paper.[170] Then fold up the bottom to meet that crease, making a crease that is one-quarter of the way up the paper.

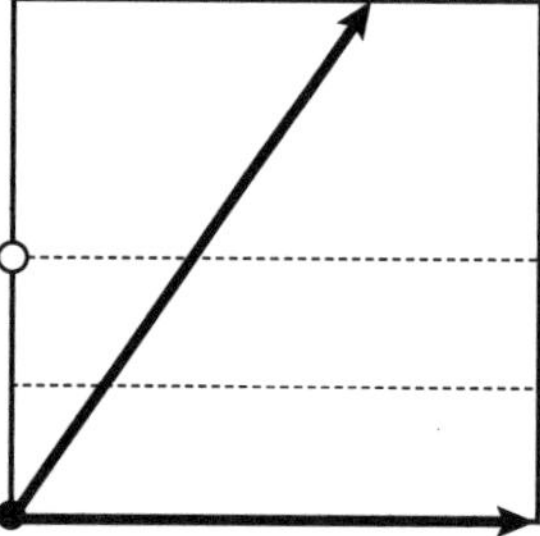

The next fold is key. Fold along a diagonal so that the white dot falls on the top arrow (or *ray*) of the angle *and* so that the black dot falls on the crease that is one-quarter of the way up.

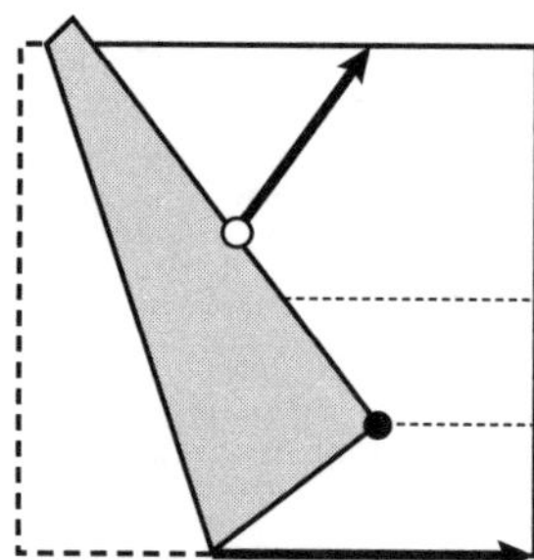

170 It actually doesn't need to be halfway up. The two rectangles at the bottom just need to be congruent.

Unfold the flap, but keep track of where the black dot fell.

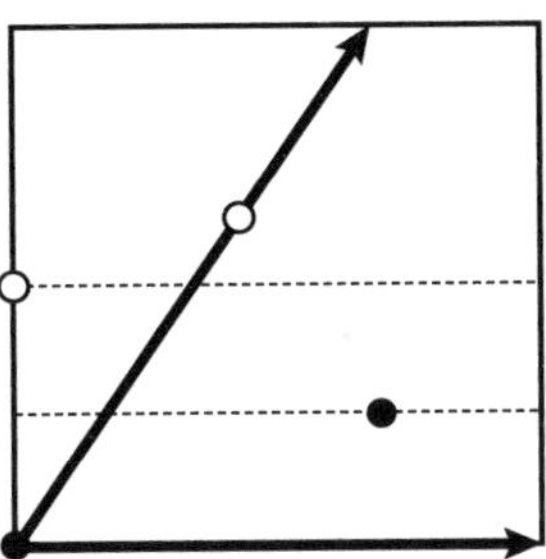

The two black dots can be connected with a ray. This can be accomplished by folding the paper, so there is no need for a straightedge. The result will be an angle that is one-third of the original angle.

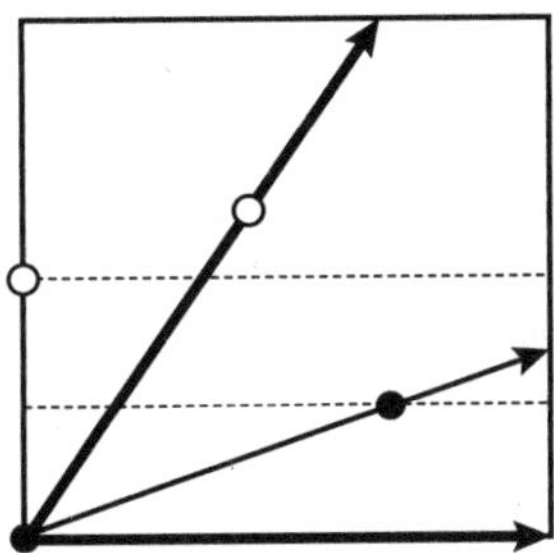

This is enough to meet the challenge of trisecting an angle, but if you want all three angles, you can fold the paper to join the top of the original angle with the newly drawn ray and unfold. The crease will complete the picture and show the original angle properly trisected.

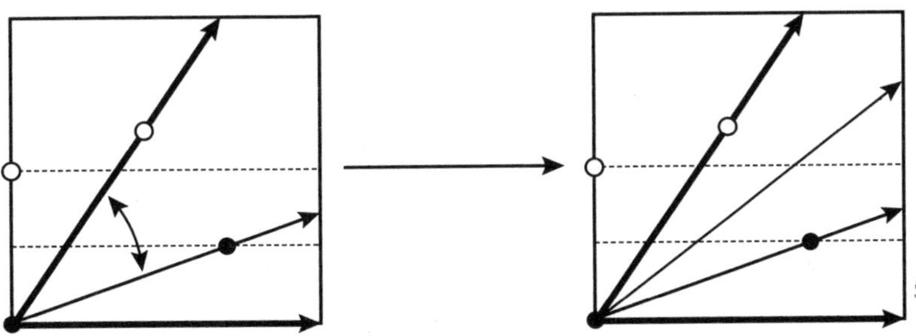

I will not give the proof for why this trisection of an angle works. If you know some things about congruent triangles, you can work out the details, and this problem is a good one for you to tackle on your own.

Paper folding accomplished what Euclidean geometry failed to do. In fact, it can be shown that paper folding can also double the cube.[171] It may be tempting to think that paper folding is more powerful than Euclidean constructions, but to be fair, that is only true in some ways. It is weaker in at least one profound way: we cannot construct circles at all using only straight folds, and therefore much of Euclidean geometry remains out of reach for mere paper folding. It might be interesting to think about a geometry that has both paper folding and a compass.

Paper folding itself, commonly known by its Japanese word *origami*, is an ancient art, but it has also interested mathematicians for a while. There is a famous problem posed independently by Grigory Margulis and Vladimir Arnold.

> *The Margulis Napkin Problem or Arnold's Ruble Problem*
> Is it possible to start with a rectangular napkin or a Russian ruble, perform a series of folds, and end up with a shape that has a greater perimeter than the original?

There is a small issue, however, with the wording of this question. The concept of "fold" is not well defined beyond the prohibition against tearing or stretching. This lack of clarity is especially problematic when the paper has multiple layers after having gone through several subsequent folds. Depending on the rules of folding, the answer to the napkin problem[172] changes.

One interpretation of a fold is highly restrictive. The paper must remain flat on the table, and every time a fold is made you must fold all layers. For example, consider a piece of paper that already has been folded once with a second fold proposed along the dotted diagonal.

171 We could also ask, "What are the folding numbers?" just like we asked, "What are the constructible numbers?" Answering this will elucidate why these two constructions are possible with paper folding but not with a straightedge and compass. As tempting as it is simply to state precisely what the folding numbers are, I will let you think about and research them.

172 I will adopt one name for the problem going forward to streamline our references, with due respect to Dr. Arnold and the Russian ruble.

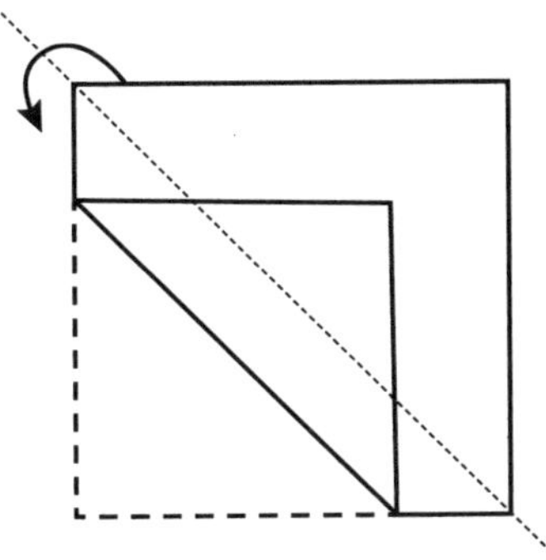

Two layers of paper are crossed by this line. In the strict interpretation of a fold, both layers must be folded.

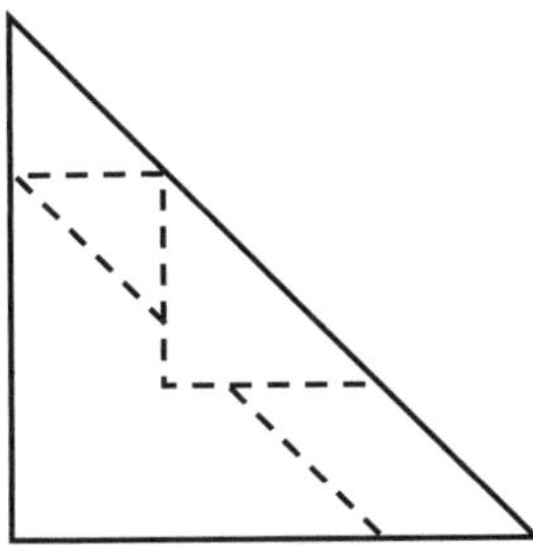

If this is what we mean by folding, then the answer to the napkin problem is no. In fact, it can be proven that every fold in this way will always decrease the perimeter, and therefore there is no way to increase beyond the original perimeter.

> *Restricted Folding Solution to the Napkin Problem*
> If a rectangular napkin is subject to flat folds that always fold every layer, then the result can never have a larger perimeter than the original napkin.

If, however, "fold" is taken to mean that *single layers* can be folded, things become more interesting. For example, if we take the same fold line but fold over only the top layer (the small triangle), we get a shape different than the previous one.

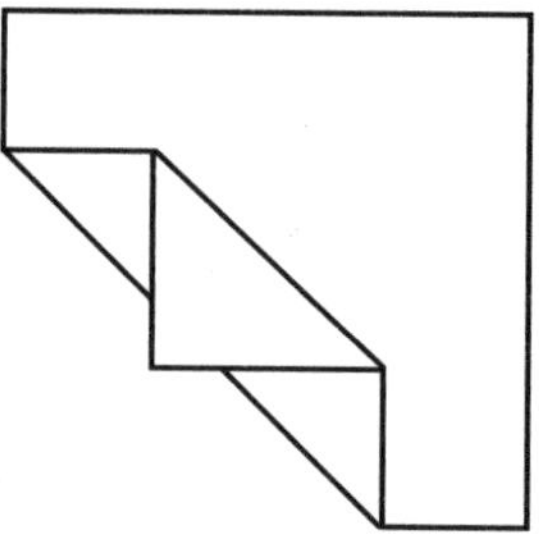

To be sure, this shape does not have a larger perimeter than the original, but it does have a perimeter that is larger than the previous step. Therefore, it at least shows that increasing the perimeter from one step to the next can be done, something that was impossible with the more restrictive definition of folding. If we only use folds that are the "folding and unfolding of single layers," it is unknown whether we can arrive at a perimeter greater than the original.

> UNSOLVED PROBLEM: Given a rectangular napkin, can a series of folding and unfolding of single layers produce a shape that has a perimeter greater than the original?

Actually, we can allow for some more complicated folds such as reverse folds, and the problem still remains unsolved. If, however, we allow the full arsenal of origami folds—still with no cutting or stretching of the paper—it *is* possible to end up with a shape that has a greater perimeter than the original square. The mathematician and origami artist Robert J. Lang showed in 1997 that this can be accomplished with some well-known origami figures, the most famous of which is the paper crane.

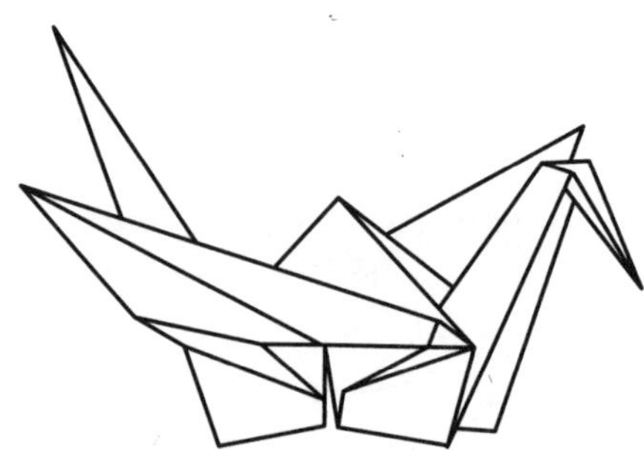

Of course, this is shape is not flat, so "perimeter" doesn't have a clear meaning. The base from which the crane comes, however, is flat. This base has four branches, two of which eventually become the wings, and the other two of which become the head and the tail. Normally the head and tail branches are "thinned out" through a fold or two, whereas the wings are not. But if we thin out all four in just the right way, we can end up with a flat shape that has a perimeter greater than the original square!

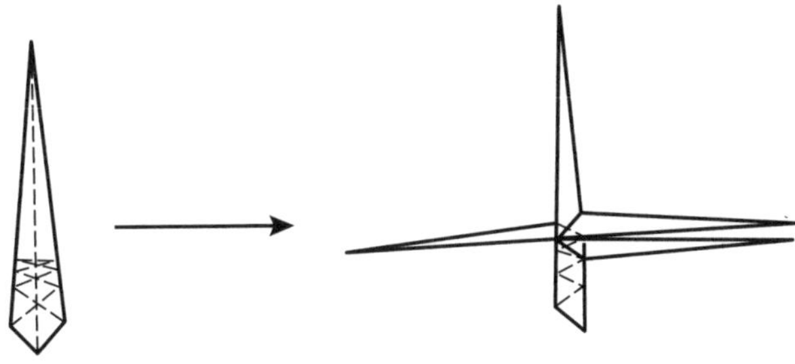

In fact, Lang showed something even more surprising: with more branches and more thinning, we can get a shape with a perimeter that is as large as we want.[173] (This, of course, supposes paper that is infinitely thin.) In other words, we can start with paper that is one inch on each side and fold it into a shape that has a perimeter of 100 miles!

> *Least Restrictive Folding Solution to the Napkin Problem*
> Using the full set of origami folds on a square piece of paper, a shape can be formed that has an arbitrarily large perimeter.

173 Lang used some folds in his proof that fall outside of *rigid origami*. Rigid origami insists that no part of the paper is temporarily "curved" to get the paper in position. In 2004, however, Alexei Tarasov proved the same theorem using only folds proper to rigid origami.

CHAPTER 3
THE TWIN PRIME CONJECTURE

EARLIER, WE spent quite a bit of time showing how the prime numbers seem to spread out as they increase. We first proved that there are arbitrarily long gaps in the primes. That is, if you want to find a sequence of one million consecutive numbers that are *not* prime, you can do it. We then played around with the remarkable Prime Number Theorem, which gave us a way to estimate how many primes exist within a particular range, and we used this to show that there are far fewer primes among larger numbers than among smaller numbers. Even though prime numbers become spread out and increasingly rare, however, they do not do so in a consistent manner. For example, while the average gap between primes less than 6,000,000,000,000 is around twenty-nine, we find immediately after this number the two primes 6,000,000,000,581 and 6,000,000,000,583, which *are only two apart!*

In fact, while it is true that there are gaps as big as we want between the primes, there is another sense in which these gaps are limited. The mathematician Joseph Bertrand (1822–1900) postulated that if you take any whole number larger than 1 and double it, you will always find a prime in between these two numbers. (Fair warning: this is not *quite* Bertrand's postulate, but I will clear that up in a minute.) For example, if you take the number 125 and double it to 250, Bertrand claimed, there will be at least one prime in between 125 and 250.[174] This may not be so surprising for small numbers, where primes seem to exist in abundance, but it is a bit more surprising for large numbers. For example, according to Bertrand we are guaranteed to have at least one prime number in between 7,000,000,000,000 and 14,000,000,000,000. In 1852, the Russian mathematician Pafnuty Lvovich Chebyshev proved Bertrand's postulate, so the theorem became known as the Bertrand–Chebyshev Theorem.

Bertrand–Chebyshev Theorem (Sort Of)
There is always a prime number between any whole number greater than 1 and twice that number. In other words, given any n greater than 1, there is always a prime between n and $2n$.

174 There are actually twenty-three of them: 127, 131, 137, 139, 149, 151, 157, 163, 167, 173, 179, 181, 191, 193, 197, 199, 211, 223, 227, 229, 233, 239, and 241.

The actual Bertrand–Chebyshev Theorem is an ever-so-slightly improved version of this formulation.

Bertrand–Chebyshev Theorem
Given any n greater than 3, there is always a prime between n and $2n-2$.

This version of the theorem is an improvement over the simpler version because it guarantees a prime number between 7,000,000,000,000 and 13,999,999,999,998 rather than between 7,000,000,000,000 and 14,000,000,000,000.[175]

We can also state the theorem in terms of when a next prime must occur. If you find a prime, say p, the next prime must be less than $2p$. We noted earlier that 6,000,000,000,583 is a prime number. Therefore, we are guaranteed that the next prime number must be no more than $2 \times$ 6,000,000,000,583, or 12,000,000,001,166.

Why does this limit on the gap between prime numbers not contradict our earlier proof that we can find gaps that are arbitrarily large? Let's look at a specific situation. The simplified version of the Bertrand–Chebyshev Theorem says that there must be a prime between 7,000,000,000,000 and 14,000,000,000,000. Our earlier proof states that we can find a sequence of 7,000,000,000,000 consecutive numbers that are *not prime*. By remembering how our earlier proof worked, we can resolve this tension. The first of our 7,000,000,000,000 consecutive non-primes was

$$7{,}000{,}000{,}000{,}001 \times 7{,}000{,}000{,}000{,}000 \times 6{,}999{,}999{,}999{,}999 \times \ldots \times 3 \times 2 + 2.$$

This number is *huge*, much larger than those in the Bertrand–Chebyshev interval of 7,000,000,000,000 to 14,000,000,000,000. In other words, the gap we constructed occurs *much* later than the number 14,000,000,000,000. The Bertrand–Chebyshev Theorem says that if you start with a number, there is a limit to how far you can go before you hit another prime, but it is still the case that you can find gaps exceeding the length of that interval if you look at significantly larger numbers.

175 The refinement to the actual theorem only provides the elimination of a single value. $2n$ is not prime, and neither is $2n-2$, because both of these values are even. Thus, the only number that is excluded as a possibility in the actual statement of the theorem versus our simplified version is $2n-1$. Nevertheless, it is still an improvement.

The French mathematician Adrien-Marie Legendre (1752–1833)[176] conjectured something similar, but to this day it remains unproven.

> *Legendre's Conjecture*
> There is always a prime number between n^2 and $(n+1)^2$ for every whole number n.

Even with Bertrand's proven limit and Legendre's conjectured limit,[177] primes still become more and more spread out as they increase—*on average*. Thus, it remains somewhat surprising that we continue to find examples such as 6,000,000,000,581 and 6,000,000,000,583, which are only two apart. Pairs of numbers such as these are called *twin primes*.

> DEFINITION: Two primes are called *twin primes* if they are exactly two apart.

There is a slew of twin prime pairs right away.

3 and 5	5 and 7	11 and 13
17 and 19	29 and 31	41 and 43

With primes becoming more and more spread out, it would make sense that these pairs become rarer and rarer. This is indeed true; they do become rarer as numbers increase. What is not clear, however, is whether they become so rare that they eventually disappear entirely. In other words, are there an infinite number of twin prime pairs? This, indeed, is one of the most famous unsolved problems in number theory.

> UNSOLVED PROBLEM: Are there a finite or an infinite number of twin prime pairs?

176 As mentioned earlier, among many of Legendre's accomplishments was a proof for the case of $n=5$ in Fermat's Last Theorem, more than 160 years before Andrew Wiles proved the general case.

177 It is worth seeing how Legendre's Conjecture relates to the Bertrand–Chebyshev Theorem. Let's take an example of $n=100$. If Legendre's Conjecture is true, then there should be a prime between 100^2 and 101^2. That is, there should be a prime between 10,000 and 10,201. The Bertrand–Chebyshev Theorem guarantees a prime between 10,000 and $2\times 10{,}000$, or 20,000. In other words, Legendre's Conjecture is trying to improve upon Bertrand's bound on the gap between primes. Bertrand's limit has been proven. Legendre's improved limit has not.

Mathematicians have not been able to prove that there are an infinite number of such pairs, but neither have they been able to prove that twin primes eventually stop occurring. There are reasonable arguments on both sides of the debate. In support of an affirmative answer, with an infinite number of primes why would we not have an infinite number of twin prime pairs? Moreover, mathematicians have looked at some *very large* numbers, and they keep finding new twin prime pairs.[178] In support of a negative answer, with primes becoming more and more spread out, it might make sense that at some point they become so spread out that primes this close together are no longer possible. Of course, this would only raise the question, "What is the *last* twin prime pair, and why do these pairs not exist after that?"

If there are an infinite number of twin prime pairs, it is a bit surprising that the proof is so difficult. After all, Euclid had little trouble showing that the number of primes was infinite. It seems like a similar proof could be used to show that the number of twin prime pairs is also infinite. Nevertheless, such a proof has yet to be found.

If it is the case that there are only a finite number of these pairs, it is still surprising that we cannot find a proof. Some related problems are not nearly as hard. For example, suppose we define a *triplet prime trio* as a set of three primes that are two apart, such as 3, 5, and 7. It is well known that there are *not* an infinite number of these triplet prime trios. In fact, the *only* one is (3, 5, 7). To explore why, let's start with two simple candidate trios. First consider the trio (137, 139, 141). Here we *almost* have a triplet prime trio. In fact, it starts with a twin prime pair: 137 and 139. Unfortunately, the third number, 141, is not prime. It is divisible by 3. Second, consider the trio (163, 165, 167). Again, we have almost succeeded. The numbers 163 and 167 are prime, but 165 is divisible by 3. In both sequences, we were foiled by a number being divisible by 3 and therefore not prime. I claim that this will always be the case. To see why, let's look at five consecutive numbers starting with a prime number, p, on a number line.

178 It is dangerous to list these sorts of things because they can become outdated, so I will simply give a very large example and resist the urge to call it the "largest known" example. The numbers $2{,}996{,}863{,}034{,}895 \times 2^{1{,}290{,}000} - 1$ and $2{,}996{,}863{,}034{,}895 \times 2^{1{,}290{,}000} + 1$ are both prime numbers and thus form a twin prime pair. They are each 388,342 digits long.

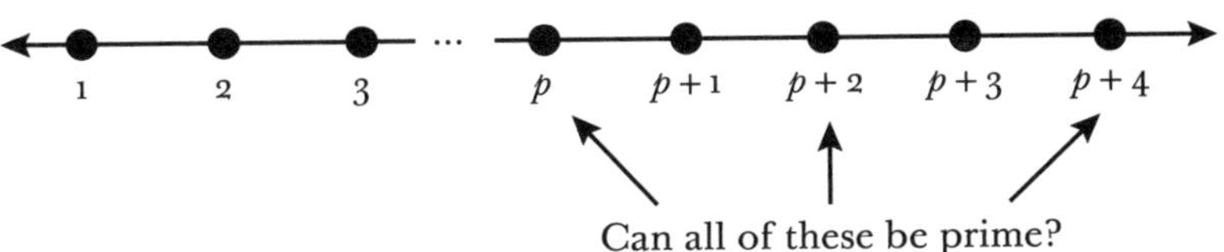

We don't need to worry about $p = 2$, because (2, 4, 6) is *not* a triplet prime trio. Nor do we need to worry about $p = 3$, because (3, 5, 7) *is* a triplet prime trio. So, for our purposes, p is a prime number further down the number line. Because multiples of 3 are always three numbers apart, one of the first three numbers starting at p must be a multiple of 3. But it cannot be p itself because the only prime number that is divisible by 3 is 3 itself.

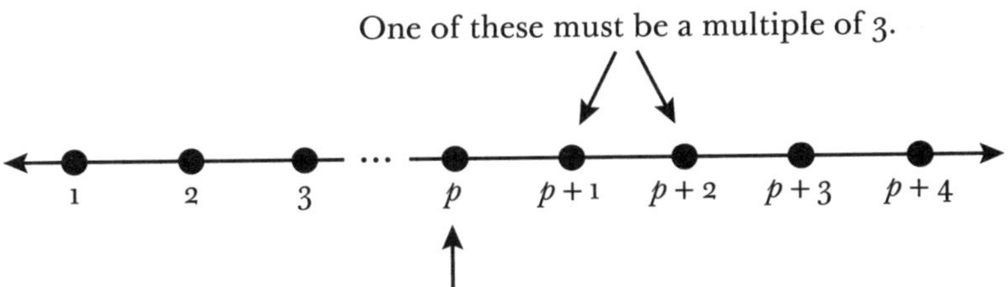

What if $p + 1$ is a multiple of 3? If we use triangles to label all multiples of 3 on our number line, we can see that $p + 4$ cannot be a prime number, and so we lose a member of our potential triplet prime trio.

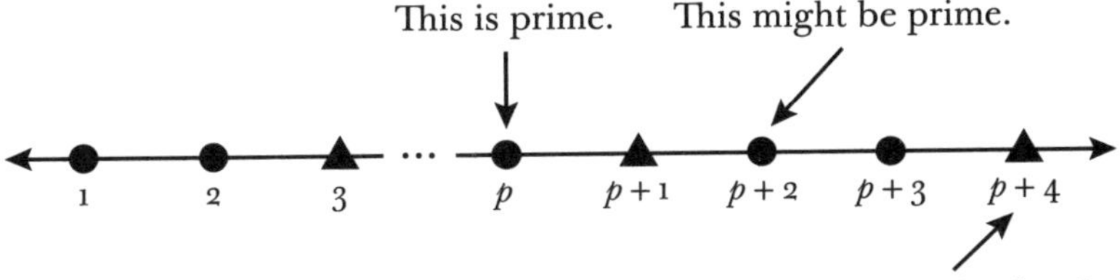

If p is not a multiple of 3, and $p + 1$ is not a multiple of 3, then the only other possibility is if $p + 2$ is a multiple of 3. In this case, however, we also lose a different member of the potential triplet prime trio.

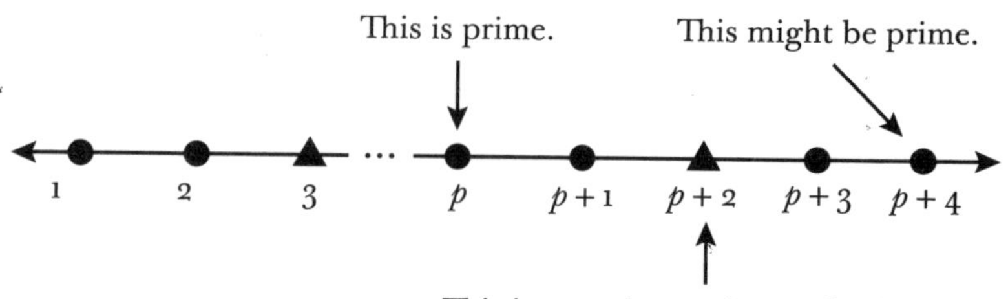

It might be helpful to summarize the argument. If p is prime, then either $p + 2$ or $p + 4$ must be a multiple of 3 because of how multiples of 3 fall on the number line. If that is the case, then $(p, p+2, p+4)$ cannot be a triplet prime trio. At least one of these numbers has to be a multiple of 3 and therefore cannot be prime. The only way around this dilemma is if $p=3$, and then we get our lone triplet prime trio, (3, 5, 7).

Our proof is complete, then. There are *not* an infinite number of triplet prime trios. There is only one. Of course, if there *had been* an infinite number of triplet prime trios, we would also have an answer to the Twin Prime Conjecture. After all, within a triple prime trio would be *two* twin prime pairs. Alas, we remain a bit mystified. On the one hand, if something went wrong with triplet prime trios such that at some point no more could exist, why is this not the case for twin prime pairs? And if it is the case, why is it so hard to prove? On the other hand, if there are an infinite number of twin prime pairs, why is it not as easy to prove as the infinitude of primes themselves? The answer to all of these questions, for now, remains a mystery.

CHAPTER 4
THE GOLDBACH CONJECTURE

BY THIS point in the book, I have made a massive deal about the fact that the prime numbers are the building blocks of multiplication. We can break up any whole number into a unique product of primes. What if there were some way that these same numbers could build up other numbers *through addition*? This is precisely what the German mathematician Christian Goldbach (1690–1764) conjectured in 1742. It is interesting that his contributions to mathematics are not well known apart from this idea: his notoriety comes almost entirely from his correspondence with Leonard Euler in which the conjecture appears. In a letter to Euler, he wrote in the margin (giving a coincidental but entertaining parallel to his French predecessor Fermat), "Every integer greater than 5 is the sum of three primes."[179] Apparently, Euler responded and reminded Goldbach that he had once claimed that this conjecture followed from a more basic one: "Every even integer greater than 2 can be written as the sum of two primes." This statement has become known as the Goldbach Conjecture.

Goldbach Conjecture
Any even integer greater than 2 can be written as the sum of two prime numbers.

Euler was, of course, correct. Goldbach's marginal note about *all* numbers is dependent upon the more basic conjecture about *even* numbers. This is because the prime 2 can be pulled off of an even number, and the prime 3 can be pulled off of an odd number—in both cases leaving an even number that the official conjecture says can be written as the sum of two primes. For example, the even number 26 can be written as $2+24$. According to the Goldbach Conjecture, 24 can then be written as the sum of two primes, namely $19+5$. Therefore, the original number, 26, is the sum of *three* primes: $2+19+5$. Similarly, the odd number 27 can be written as $3+24$, and so is also a sum of three primes: $3+19+5$.[180] Of

179 Actually, his conjecture read, "Every integer greater than 2 can be written as a sum of three primes," but this is because he considered 1 to be a prime number. This is not true by modern convention, as I have noted. Therefore, I have updated the Goldbach Conjecture accordingly. I am not alone: the formulation I have given is consistent with how most mathematicians would state it.

180 For a more formal proof, take any integer, n, that is greater than 5. If n is even, then $n-2$ must be even. If th Goldbach Conjecture is correct, $n-2$ can be written as the sum of two primes, p and q. Therefore, $n=(n-2)+2$,

course, this all depends on the Goldbach Conjecture's being true. To this, Euler responded, "I think this is a fully certain theorem, although I cannot demonstrate it." To this day, it remains unproven.

In exploring any conjecture, it is worth taking the time to list several examples. Let's pick an even number, say 16. There are eight ways of breaking the number 16 apart into a sum. (Think back to our addition trees—there are eight ways to start the tree for 16.)

1+15	2+14	3+13	4+12
5+11	6+10	7+9	8+8

Do any of these sums consist of two prime numbers? In fact, there are *two* such sums: 3+13 and 5+11. Therefore, the conjecture is unsurprisingly confirmed, at least for the number 16.

We can proceed in one of two directions at this point. We could start by listing numbers and finding ways to write them as the sum of two primes, or we could start by adding the combinations of two primes together to see which numbers are produced. We are going to take the second path. Let's list the primes across the top of a grid and again down the left side to make an "addition chart" of prime numbers, which we can then fill in with the appropriate sums.

+	2	3	5	7	11	13	17	19	23
2	4	5	7	9	13	15	19	21	25
3	5	6	8	10	14	16	20	22	26
5	7	8	10	12	16	18	22	24	28
7	9	10	12	14	18	20	24	26	30
11	13	14	16	18	22	24	28	30	34
13	15	16	18	20	24	26	30	32	36
17	19	20	22	24	28	30	34	36	40
19	21	22	24	26	30	32	36	38	42
23	25	26	28	30	34	36	40	42	46

or $n=p+q+2$, making it the sum of three primes. If n is odd, then $n-3$ must be even. Again, if the Goldbach Conjecture is correct, $n-3$ can be written as the sum of two primes, p and q. Therefore $n=(n-3)+3$, or $n=p+q+3$, making it again the sum of three primes.

There are a couple of things to notice about the table. In the first row and the first column, all of the numbers except for 4 are odd. This makes sense because these are the only entries that involve the number 2. All primes except for 2 are odd. With the exception of $2+2$, every other sum involving 2 must be 2 plus an odd number, which always result in another odd number. In every other case, the sum of two primes is even. Although the Goldbach Conjecture deals only with even numbers, we still include the first row and column because their intersection is the only cell that contains the number 4, which is important.

The next observation is the symmetry in the table. Look at the diagonal joining the two 17s: 15, 14, 12, 12, 14, 15. It is symmetric with the two 12s in the middle. An identical symmetry will occur in any addition table that uses the same numbers across as it does down. The entries simply swap the order of addition. For example, in the 17-diagonal, the entry for one of the 14s is the result of $11+3$, and the entry for the other 14 is the result of $3+11$. The practical ramification of this symmetry is that we only have to focus on half of the table.

At this point, we should remind ourselves of the goal. We are looking for even numbers, or rather we are looking for whether any even numbers seem to be missing. Goldbach, after all, conjectures that there won't be any missing even numbers. I will not shade all occurrences, but a quick glance suggests that we get all of the even numbers simply by moving more or less down the main diagonal.

+	2	3	5	7	11	13	17	19	23
2	4	5	7	9	13	15	19	21	25
3	5	6	8	10	14	16	20	22	26
5	7	8	10	12	16	18	22	24	28
7	9	10	12	14	18	20	24	26	30
11	13	14	16	18	22	24	28	30	34
13	15	16	18	20	24	26	30	32	36
17	19	20	22	24	28	30	34	36	40
19	21	22	24	26	30	32	36	38	42
23	25	26	28	30	34	36	40	42	46

We might think that this observation could lead to a proof that all even numbers eventually will exist along this diagonal, and yet the conjecture remains unproven. There are two reasons why this proof might be difficult. The first is perhaps more obvious. Prime numbers are the building blocks of multiplication, but here we are trying to combine them using addition. We know lots of properties of prime numbers that have to do with multiplication and division, but combining them with addition is a harder nut to crack. That said, if the Goldbach Conjecture is true—and most mathematicians think it is—then this connection between primes and addition is rather incredible.

The second reason the conjecture might be hard to prove is a bit counterintuitive. We began by combining primes in our table to see which even numbers arise, but if we go the other way and start by decomposing an even number, we find something interesting. For example, we can write the number 22 as the sum of primes in *three different ways*: 19 + 3, 17 + 5, and 11 + 11. (This does not include the symmetric versions of 3 + 19 and 5 + 17.) Some numbers have even more unique decompositions. For example, the number 100 can be written as the sum of two primes in *six* different ways.

3 + 97	11 + 89	17 + 83
29 + 71	41 + 59	47 + 53

I will not list them all, but the number 120 can be written in *twelve* different ways. In fact, it is generally true that the larger the number, the more ways we can find to write it as the sum of primes. By "generally true," I mean that the number of ways doesn't increase with *every* number. For example, even though 120 has twelve ways, 122 has only four. As a general pattern, however, larger numbers have a greater number of possible decompositions. For some very large even numbers, there are a lot of ways to write them as the sum of two primes. In other words, the Goldbach Conjecture seems to be not only true but *very* true, especially for large numbers. The fact that the numbers can be decomposed in so many ways could be preventing us from finding a proof. If there were a *unique* way to write every even number as the sum of two primes, then we might be able to find a pattern or even a formula for those two primes. With a formula, we might be on our way to a proof. Sometimes an abundance of riches causes complexity in finding a path forward.

We close this chapter with a reminder that the "official" Goldbach Conjecture is about even numbers, but if that conjecture is true, so too is his original note to Euler: "*Every* integer greater than 5 is the sum of three primes." That would mean that the primes not only are the building blocks of multiplication but also provide a set of building blocks of addition from which we can get the whole numbers. What mysterious things these prime numbers seem to be.

CHAPTER 5
THE QUEST FOR A FORMULA THAT GENERATES PRIME NUMBERS

PRIME NUMBERS have played an important role in every stage of our journey thus far. There are good reasons for this. To start with, they are fundamental objects for arithmetic and are infinitely interesting. This is not the last chapter of the book, but it is the last one that deals with prime numbers. Perhaps one of the most important open problems in the mathematics of primes is to develop a simple and efficient formula that produces them. For example, consider the expression n^2+n+41. Let's use this formula to calculate values starting at $n=0$.

$$0^2+0+41=41$$
$$1^2+1+41=43$$
$$2^2+2+41=47$$
$$3^2+3+41=53$$
$$4^2+4+41=61$$

Notice that all of these numbers are prime. In fact, this remarkable formula continues to produce prime numbers all the way through $n=40$.[181] The problem comes when $n=41$. Without calculating the monstrously large value, can we convince ourselves that $41^2+41+41$ cannot possibly be prime? Multiples of 41 are exactly 41 apart. Moreover, 41^2 is a multiple of 41, being 41×41. Therefore, the next multiple of 41 after 41^2 is 41^2+41, and the one after that is exactly $41^2+41+41$. As a multiple of 41, this number cannot be prime.

It is too bad, actually. If a formula like this existed, we could use it to help answer questions like the Twin Prime Conjecture.[182] Without a formula, it may be hard to prove that there are an infinite number of twin prime pairs, but once we have one, we have the full power of algebra at our disposal. We could look to see whether this formula can produce an infinite number of values that are exactly two apart. Of course, that would depend on the formula's not only producing primes but also producing

181 By this point, it should not surprise us to learn that Euler is the one credited with finding this remarkable formula.

182 The formula $224{,}584{,}605{,}939{,}537{,}911+18{,}135{,}696{,}597{,}948{,}930n$ also produces prime numbers for the first twenty-seven values starting at $n=0$. This is perhaps more amazing because it is a simple linear equation, but perhaps less so in that the primes it produces are huge.

all of the primes.[183] If we could do this, it would be a tremendous advancement of our knowledge of prime numbers. While such a formula might exist, mathematicians have proven that it cannot be a *polynomial* equation. Given this disqualification of polynomials—which are in some way the archetype of simple formulas—I should stress that the type of formula we are looking for would need to be easy enough to manipulate algebraically and efficient enough to produce primes without much of a computational headache. Otherwise, it wouldn't be useful. In fact, we know of at least two interesting albeit non-useful formulas that *do* produce the prime numbers.

Arguably, the first "formula" is more of a process.

1. Start with a number, n.
2. Calculate $(n-1) \times (n-2) \times \ldots \times 3 \times 2$.
3. Calculate the remainder of step 2 when divided by n.
4. Divide step 3 by $n-1$, rounding down if necessary.
5. Multiply step 4 by $n-2$.
6. Add 2 to the result.

It is helpful to look at an example. Let's take the number $n=7$.

1. Start with $n=7$.
2. $6 \times 5 \times 4 \times 3 \times 2 = 720$
3. When 720 is divided by 7, the answer is 102, remainder 6.
4. $6 \div (7-1) = 1$
5. $1 \times (7-2) = 5$
6. $5 + 2 = 7$

The result is 7, which is a prime number. Of course, this is the number with which we started, so let's look at a second example that starts with $n=8$.

1. Start with $n=8$.
2. $7 \times 6 \times 5 \times 4 \times 3 \times 2 = 5{,}040$
3. When 5,040 is divided by 8, the answer is 630, remainder 0.
4. $0 \div (8-1) = 0$

183 Our simple polynomial formula not only misses primes less than 41 but also skips the prime 59, which is particularly relevant for the Twin Prime Conjecture because of its pairing with 61.

5. $0 \times (8 - 2) = 0$
6. $0 + 2 = 2$

This time the result is 2, which is still a prime number. Let's now use this "formula" to calculate more values.

Starting Number	Result
2	2
3	3
4	2
5	5
6	2
7	7
8	2
9	2

Starting Number	Result
10	2
11	11
12	2
13	13
14	2
15	2
16	2
17	17

Notice that most of the results are 2. In fact, the result is only different from 2 when n is a prime number, in which case the result is precisely that prime number. In other words, this fascinating and somewhat complicated process accomplishes a rather simple task.

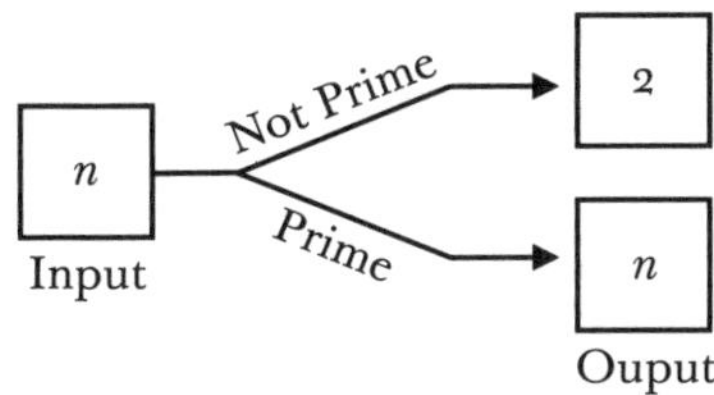

That said, in every case the formula produces a prime, and moreover it produces *all* of the primes. In some ways, though, it cheats. It takes every non-prime and maps it to 2, and every prime and maps it to itself. But this shouldn't take anything away from the elegance of the process. It is phenomenal that it can even isolate the primes and leave them unchanged, all the while taking non-primes and turning them into 2s.[184] That said, the problem with using this process to prove things about primes is threefold. First, it is not exactly algebraically friendly.[185] Second, it is a computational disaster. Just to get the prime 23, we have to start by calculating $23 \times 22 \times 21 \times 20 \times 19 \times 18 \times 17 \times 16 \times 15 \times 14 \times 13 \times 12 \times 11 \times 10 \times 9 \times 8 \times 7 \times 6 \times 5 \times 4 \times 3 \times 2$. Third, and maybe most importantly, in order to know whether the formula produces a prime, we have to know whether the input is prime, which sort of defeats the purpose!

The second example of a prime-producing formula is recursive. We first encountered the concept of a recursive formula in the chapter on the Fibonacci numbers. In our new recursive formula, the first value is

$$p_1 = 2.920050977316134712092562917112019\ldots.$$

After, that, the other values are defined recursively.

$$p_n = (\text{the whole number part of } p_{n-1}) \times (\text{the decimal part of } p_{n-1} + 1)$$

The results of this formula are not whole numbers, but if we round them down, it turns out that we get the prime numbers.

Wait—I skipped over something. The value of p_1 is what? And what are the other decimals after the ellipses? Well, let's see how the formula works, and then we will come back to these questions. We start by generating the first prime. It should be the rounded-down version of p_1, which is 2. To get the next prime, we first need to calculate p_2.

$$p_2 = (\text{the whole number part of } p_1) \times (\text{the decimal part of } p_1 + 1), \text{ or}$$
$$p_2 = 2 \times (0.92005\ldots + 1) = 3.8401\ldots$$

184 I will sketch out why this process works. The key step is the second one. If n is not prime, then it has factors, and the product $n!$ explicitly will contain those factors. At least two of these factors will multiply together to be n itself. For example, when $n=8$, we find both 2 and 4 in $8!$, and $2\times4=8$. This means that the number in the second step is divisible by n when n is not prime. Therefore, the remainder in the third step is 0, and it won't take long to convince yourself that this will result in a 2 in the final step. The argument does not work for the number $n=4$ because it is the only composite number that fails to have a matching pair of distinct factors. But you can trace the number 4 through the process to see that it still ends in 2. If n is a prime number, then there are no factors of n in the second step. This is a much harder argument, but a well-known result in number theory is that when n is a prime number, the remainder of $(n-1)!$ when divided by n is $n-1$. For example, the remainder of $4!$ when divided by 5 is 4. If the number in step two is $n-1$, it won't take long to convince yourself that this will result in n in the final step.

185 There is an algebraic formula for this process: $\left\lfloor \frac{(n-1)!\,\%\,n}{n-1} \right\rfloor \times (n-2) + 2$, where "%" means "the remainder when divided by" (formally *modulo*), and the $\lfloor\_\rfloor$ symbol means "round down." Even in this form, however, this is not algebraically friendly.

We then round 3.8401... down to 3, which is the next prime number after 2. We will do just one more calculation.

$$p_3 = (\text{the whole number part of } p_2) \times (\text{the decimal part of } p_2 + 1), \text{ or}$$
$$p_3 = 3 \times (0.8401\ldots + 1) = 5.520\ldots$$

Rounded down, 5.520... is 5, which is the next prime! It is rather astounding that this process works: it produces not just primes but all the primes, and in order. But it raises a major question: "What *is* this mysterious number 2.920050977316134712092562917112019..., and *what are rest of the digits?*"

Therein lies the first problem. The recursive formula depends on knowing the digits of this important number, and the more digits we know, the more primes we can generate. The above listing has enough digits to generate at least twenty-five primes. If this magic number is so important, how do we get more digits so that we can produce more primes? Therein lies the second problem. It turns out that the value of p_1 is an infinite sum.

$$p_1 = \frac{2-1}{1} + \frac{3-1}{2} + \frac{5-1}{2\times3} + \frac{7-1}{2\times3\times5} + \frac{11-1}{2\times3\times5\times7} + \ldots$$

The fact that the value of p_1 is an infinite sum is not an issue; we saw several infinite sums in our explorations of π and e. It is the particulars of this sum that are the concern. The numerators are all of the form "prime -1," and the denominators are all products of primes. Therefore, to generate primes, we first need this number, and to find it we need a list of primes. It is tragically circular. All hope may not be lost, however: maybe we can discover more about this mysterious number. Maybe someday we can find another representation of it that does not depend on prime numbers. After all, π had more than one infinite representation. If we can find such a patterned representation, then we may be on to something. Until then, the quest for a convenient prime-producing formula continues.

UNSOLVED PROBLEM: Find a simple and efficient formula that produces all of the prime numbers.

CHAPTER 6
ARE THERE ANY ODD PERFECT NUMBERS?

PRIME NUMBERS are, shall we say, rather *deficient* in their divisors. In fact, they only have two: 1 and the number itself. There are other numbers that are rather *abundant* in their divisors. Think of the number 60, which has no less than *twelve* divisors: 1, 2, 3, 4, 5, 6, 10, 12, 15, 20, 30, and 60. This plentitude of divisors is remarkable especially given the relatively small size of the number 60. We expect large numbers to have many divisors, or at least we expect them to have a greater chance of having many divisors. To see twelve divisors in a number as small as 60 is more notable. In fact, 60 has more divisors than any number that comes before it, and it will not be beat until the number 120, which has just four more at sixteen divisors.

There is a (relatively) easy way to calculate how many divisors a number has using its unique factorization. Take the number 60, which is equal to $2^2 \times 3 \times 5$. Any divisor of 60 must share these prime factors and cannot have others. For example, 2×3 is a divisor of 60, as is $2^2 \times 5$. We can use this prime factorization to construct what divisors of 60 look like. They all must have the form $2^{\blacksquare} \times 3^{\blacksquare} \times 5^{\blacksquare}$. The black boxes must have the same exponents *or fewer* when compared with their counterpart in the original factorization. In other words, the black box over the 2 must be 0, 1, or 2, because the exponent of 2 in the original factorization was 2. The black boxes over 3 and 5 must be either 0 or 1. Note that 0 represents choosing no copies of that prime factor for our divisor. It also represents the reality that $2^0 = 1$. To help clarify all of this, let's look at how some of the factors of 60 result from "choosing" the exponents represented in the black box.

Factor	Choices	Prime Factorization
30	one 2, one 3, and one 5	$2^1 \times 3^1 \times 5^1$
20	two 2s, no 3s, and one 5	$2^2 \times 3^0 \times 5^1$
3	no 2s, one 3, and no 5s	$2^0 \times 3^1 \times 5^0$
1	no 2s, no 3s, and no 5s	$2^0 \times 3^0 \times 5^0$
60	two 2s, one 3, and one 5	$2^2 \times 3^1 \times 5^1$

With three choices (the numbers 0, 1, and 2) for the black box exponent of 2, two choices (0 and 1) for the black box exponent of 3, and two choices (0 or 1) for the black box exponent of 5, there are a total of $3 \times 2 \times 2 = 12$ possible factors, which is exactly what we found when we listed out all of the factors.[186]

This technique is particularly helpful in calculating the number of divisors for large numbers. For example, if we wanted to know how many divisors 15,400 has, we would start with the prime factorization: $15{,}400 = 2^3 \times 5^2 \times 7 \times 11$. As such, a divisor of 15,400 must look like $2^{\blacksquare} \times 5^{\blacksquare} \times 7^{\blacksquare} \times 11^{\blacksquare}$. The black box exponent for 2 must be 0, 1, 2, or 3 (four choices). The black box exponent for 5 must be 0, 1, or 2 (three choices). The black box exponents for 7 and 11 can only be 0 or 1 (two choices each). Therefore, the total number of possible divisors for 15,400 is $4 \times 3 \times 2 \times 2 = 48$.

One of the ways mathematicians have chosen to think about how the divisors of a number compare with the number itself is to calculate their sum. For example, the divisors of 60 are 1, 2, 3, 4, 5, 6, 10, 12, 15, 20, 30, and 60, so their sum is $1+2+3+4+5+6+10+12+15+20+30+60=168$. The accepted cutoff for having *many* divisors versus *few* divisors is whether or not this sum exceeds twice the original number. The number 60 is called *abundant* because the sum of its divisors is 168, which is greater than 2×60, or 120. The number 15 is of a different sort. It only has four divisors (1, 3, 5, and 15), the sum of which is $1+3+5+15=24$. This sum, 24, is less than 2×15, or 30. Such numbers are called *deficient*. Then there are the numbers that are "just right." Their divisors add to *exactly* twice the number. For example, the number 28 has divisors 1, 2, 4, 7, 14, and 28. The sum of these divisors is $1+2+4+7+14+28=56$, which is 2×28. These numbers are called *perfect*.[187]

Actually, the terms *abundant, deficient,* and *perfect* are usually defined a bit differently. They normally refer only to a number's *proper* divisors, which do not include the number itself. For example, the divisors of 28

186 Whenever we have several categories of choices, the total number of possibilities can be found by multiplying the number of possible choices in each category together. For example, if I have three colors of socks (red, blue, and green) and two colors of shoes (black and brown), there are $2 \times 3 = 6$ total combinations. This is because each choice of sock has two possible shoe pairings, i.e., (red socks, black shoes) and (red socks, brown shoes). It is essentially three groups of 2.

187 One particularly elegant property of perfect numbers is that the reciprocals of their factors (including the number itself) always add up to 2. For example, for the perfect number 28, the sum of the reciprocals is $\frac{1}{1}+\frac{1}{2}+\frac{1}{4}+\frac{1}{7}+\frac{1}{14}+\frac{1}{28}=2$. This is easy to prove. Because the divisors, by definition, have to add to twice the number, it must be that $1+2+4+7+14+28=2 \times 28$. Divide both sides of this equation by 28. This argument will work with any perfect number.

are 1, 2, 4, 7, 14, and 28, but the *proper* divisors are only 1, 2, 4, 7, and 14. Using this language, a perfect number is one for which the proper divisors add up to exactly the original number. In the case of the number 28, the proper divisors add up to $1+2+4+7+14=28$. This definition is admittedly better. "Adding to twice the number" seems a bit clunky in the first definition, but "adding to *exactly* the number itself" is quite elegant.

DEFINITION: If the sum of the proper divisors of a number equals the number, then it is a *perfect number*. If the sum is greater than the number, it is an *abundant number*. If the sum is less than the number, it is a *deficient number*.

Euclid himself knew about perfect numbers. In Book VII of the *Elements*, he defined them as we have done but with slightly different language. Euclid says that a perfect number is "a number that is the sum of its own parts," where "parts" means the proper factors. In typical Euclidean form, his definition is the most elegant of all.

The number 28 is not the first perfect number. That honor goes to the number 6. The proper divisors of 6, added together, are $1+2+3=6$. Perfection. Coming up with perfect numbers is something that interested Euclid. In fact, in the very last of his number theory propositions, found in Book IX, he shows how to construct perfect numbers. As the last in this category of propositions, it is something of the culmination of his work in number theory.

> *If as many numbers as we please beginning from a unit be set out continuously in double proportions, until the sum of all becomes prime, and if the sum multiplied into the last make some number, the product will be perfect.*

It is a little tricky to translate this into modern mathematical parlance, but we will give it a go. Euclid starts with the number 1 (the "unit") and says to double it as many times as we want: 1, 2, 4, 8, 16, ... The result is a list containing consecutive powers of 2. At whatever point we stop listing powers of 2, we then add up the numbers in the list: $1+2+4+8+16=31$. Check to see that 31 is prime, which it is. Finally, we multiply this sum by "the last"—meaning the last power of 2 in the list we used—which for us was 16. The result, $31 \times 16 = 496$, should be a perfect number. We can check this by listing and adding together the proper divisors of 496: $1+2+4+8+16+31+62+124+248=496$.

The two perfect numbers we found earlier, 6 and 28, can be constructed with this process, and we can use Euclid's method to find more perfect numbers.

Doubles	Sum of Doubles	Is the sum prime?	Sum × Last Double *(Perfect Number)*
1, 2	$1+2=3$	3 is prime.	$3 \times 2 = 6$
1, 2, 4	$1+2+4=7$	7 is prime.	$7 \times 4 = 28$
1, 2, 4, 8	$1+2+4+8=15$	15 is *not* prime.	N/A
1, 2, 4, 8, 16	$1+2+4+8+16=31$	31 is prime.	$31 \times 16 = 496$
1, 2, 4, 8, 16, 32	$1+2+4+8+16+32=63$	63 is *not* prime.	N/A
1, 2, 4, 8, 16, 32, 64	$1+2+4+8+16+32+64=127$	127 is prime.	$127 \times 64 = 8{,}128$

The primes that show up in the "Sum of Doubles" column are known as *Mersenne primes*, named after Marin Mersenne, the French friar who studied them.[188] Euclid was able to prove that this process will always generate perfect numbers, so long as there is a yes in the "Is the sum prime?" column. Two thousand years later, Euler proved that this process will lead to *all* of the *even* perfect numbers. The combination of these two results is known as the Euclid–Euler Theorem.[189]

> *The Euclid–Euler Theorem*
> When consecutive powers of 2 are added together to get a prime number, the prime number multiplied by the last power of 2 will always be an even perfect number. Moreover, this process will produce *all the even perfect numbers.*

188 It is worth noting that the numbers in this column, which are the sums of consecutive powers of 2, end up being one less than the next power of 2. This can be shown in a very elegant way by writing the numbers in binary form, but I leave it to you to work this out. Nevertheless, for this reason, the standard definition of a Mersenne prime is "a prime in the form of 2^p-1." To be clear, not all numbers in the form of 2^p-1 are prime, as we can see in the case of $2^4-1=15$. In fact, if p is not prime, we can prove the 2^p-1 is never prime. In other words, for 2^p-1 to be prime, p itself must be prime, but we also know that this is not enough. There are even cases where p is prime, but 2^p-1 is not prime.

189 Using the algebraic representation from the previous footnote, the Euclid–Euler Theorem states that an even number is perfect if and only if it is of the form $2^{p-1}(2^p-1)$, where 2^p-1 is a prime number.

There are two unsolved problems related to Euclid's perfection-producing process. First, we do not know whether there are an infinite number of even perfect numbers. Why not? Can't we just run Euclid's process with higher and higher powers of 2 to produce more and more perfect numbers? The problem is that the production of a perfect number via Euclid depends on the sum of doubles' being a prime number. This is the second unsolved problem: we do not know whether sums of consecutive powers of 2 will produce an infinite number of primes. After all, they don't *always* produce a prime. The numbers 15 and 63 show up early in the table, and neither of them is prime. It could be that after some point this process stops producing primes altogether. Even though there are an infinite number of primes, we do not know whether there are an infinite number of Mersenne primes. The unknown infinitude of Mersenne primes is directly related to the unknown infinitude of even perfect numbers.

UNSOLVED PROBLEM: Are there an infinite number of Mersenne primes?

UNSOLVED PROBLEM: Are there an infinite number of even perfect numbers?

But the question of how many even perfect numbers exist is not the most famous unsolved problem about perfect numbers. Notice that all of the perfect numbers resulting from Euclid's process have one thing in common. They are *even*. In fact, the only perfect numbers we know of are even. We do not know whether any odd ones exist. We have not been able to find any, but neither do we have a proof that they do *not* exist. If there are no odd perfect numbers, it seems like we should be able to identify some fatal flaw that prevents these two properties from coexisting in the same number. To date, however, no such flaw has been found, so the proof remains elusive. Of course, it could be that there is no flaw and that somewhere out there, beyond the extremely large numbers we have checked, we will find an odd perfect number. If so, it won't be Euclid's process that produces it. His algorithm can never generate an odd number because the last step is to multiply by a power of 2. Thus, if odd perfect numbers exist, we will need another way to find them. If we do discover one, it would raise the question, "Why do the numbers need to become so large before we get an odd perfect number?" The whole thing seems very strange, one might even say "perfectly odd."

UNSOLVED PROBLEM: Are there any odd perfect numbers?

CHAPTER 7
THE COLLATZ CONJECTURE

SOME MATHEMATICAL problems are known for curious reasons, and the depth of these problems may not be immediately clear. The Collatz Conjecture is one such example. In one sense, it is just plain fun. In another sense, who knows what applications may come from it or what mathematics may be developed in the quest to solve it? The problem begins with something that feels like a delightful way to practice math facts in an elementary school class.

1. Start with any positive whole number.
2. If it is even, divide the number by 2. If it is odd, multiply the number by 3 and add 1.
3. Rinse and repeat.

Let's get used to the process by considering some examples. What happens if we start with the number 5?

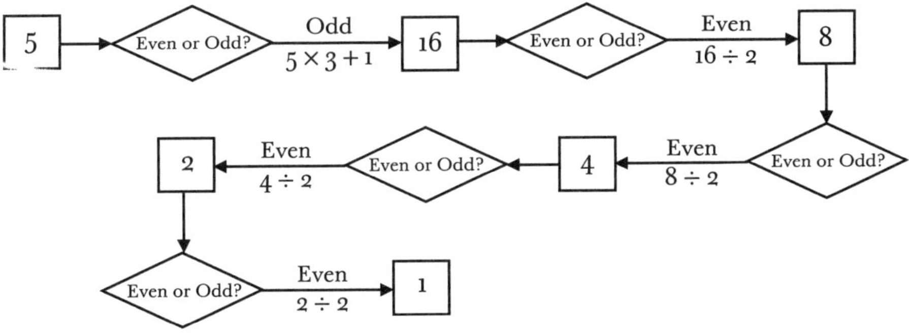

The process ends at 1.[190] Let's try starting with a different number, say 7.

190 We could ask, "What happens if we continue the process once we get to 1?" It will not take you long to realize that you get stuck in a loop: $1 \to 4 \to 2 \to 1 \to \cdots$.

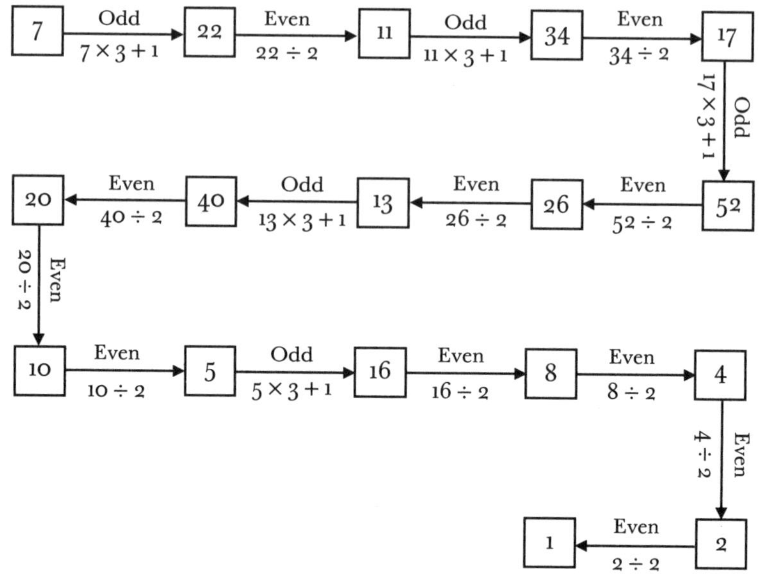

The process still ends at 1. In fact, we could have cut our work short by just a bit. Once we hit 5, we knew it would arrive at 1 from our first example. The Collatz Conjecture says that this *will always happen*. This conjecture is named after the German mathematician Lothar Collatz, who posed the problem in 1927.

> *Collatz Conjecture*
> Start with any positive integer. If it is even, divide the number by 2. If it is odd, multiply the number by 3 and add 1. Repeat this process with the new number. Eventually, the repeated process will always land on the number 1.

When we first see this process, it is tempting to say, "Of course it will end. Eventually we will end up dividing by 2 over and over again, so the sequence of numbers will become smaller and end up at 1." This argument is not unreasonable, but it is far from convincing. One obvious reason is that while "divide by 2" makes our numbers smaller, "multiply by 3 and add 1" makes them larger. To see a little more clearly what is going on, it is helpful to redraw the sequence that starts at 7 but indicate in our drawing when the numbers go up and when the numbers go down.

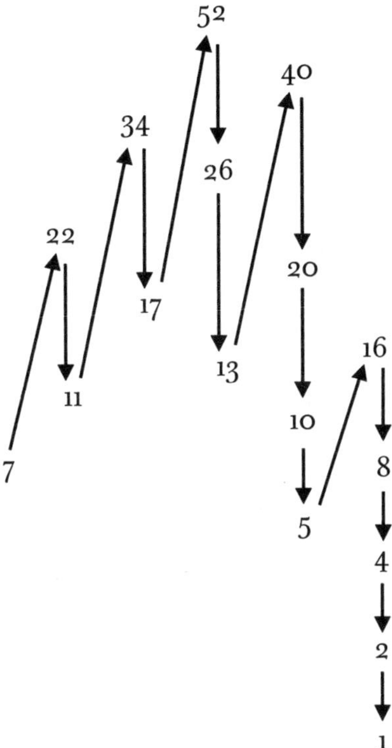

While it is true that the numbers in this example eventually fall down to 1, it is equally true that they go through a fair amount of increase along the way. Because of this up-and-down motion, numbers in a Collatz sequence are sometimes called *hailstone numbers*. Hailstones were thought to be formed by a cycle of falling and rising, and with each cycle the ice balls pick up moisture and refreeze. Eventually gravity takes over and the hailstones fall to the earth. With this analogy, we might be able to make the argument, "Everything that goes up must come down—all the way down."[191] Is there a "mathematical gravity" in the Collatz process that will eventually make all numbers fall to 1?

There are actually two ways that a number might never make it to 1.

191 It is true that the physics of hailstones is considerably more complicated. We now know that the formation of hail has to do with the properties of supercooled water droplets. Nevertheless, the analogy is sufficient for our purposes. As we extend the analogy going forward, the physics become less and less accurate, but we will continue to find it helpful for describing what might go on with these numbers. I beg the reader's indulgence, especially if he or she is a meteorologist or a physicist.

One is that the multiplicative "updraft" becomes too powerful for the division "gravity" to bring it back down. The hailstone analogy breaks down a little here, but it might be something like an "updraft" launching the hailstone higher and higher with each gust so that it eventually breaks the gravitational pull of the Earth and goes off into outer space. Mathematically, it would mean that, even with an up-and-down motion, the numbers eventually go more up than down, so overall the sequence eventually heads off towards infinity. After all, "multiply by 3 and add 1" seems to increase a number more than "divide by 2" decreases it. For example, when 17 moved up to 52, it was an increase of 35, whereas the move from 52 down to 26 was only a decrease of 26. Said differently, the number went up by more than a factor of 3 and down by only a factor of 2. Of course, the problem with "multiply by 3 and add 1" is that it can never occur twice in a row. This is because the operation only kicks in on odd numbers. An odd number multiplied by 3 is again odd, and adding 1 to it makes it even. As an even number, it would then go through the "divide by 2" part of the process. On the other hand, the "divide by 2" part *can* occur twice in a row, as in the case of 52→26→13. In fact, in moving from 16, "divide by 2" occurred *four* times in a row. These successive "divide by 2" steps happen when a number is "very" even, meaning it has a lot of factors of 2 in it.

The "Collatz gravity effect" takes over completely when the number is a power of 2, at which point the sequence plunges down to 1 very quickly. This is what happened with 16, or 2^4. When dividing by 2, the sequence traveled down through all of the other powers of 2, finally stopping at 1: 16→8→4→2→1. The whole question behind the Collatz Conjecture is how effective the "multiply by 3 and add 1" operation is at avoiding powers of 2. If it cannot avoid them indefinitely, then the conjecture would be true.

But there is a second way that a number might never reach 1. Maybe in the process of going up and down, a number comes back down to where it started. This type of sequence would create an infinite loop, as if the hailstone were caught in a perpetual cycle of updrafts and downdrafts, neither breaking the force of gravity nor being heavy enough to fall to the ground.

In fact, small variations of the Collatz process produce both of the above phenomena: going off towards infinity and getting caught in a loop. For example, suppose the odd numbers move by way of "multiply by 3 and *subtract* 1," with even numbers still moving according to "divide by 2." If we start with the number 7, which produced a lengthy but eventually terminating sequence in the original Collatz process, we get a loop.

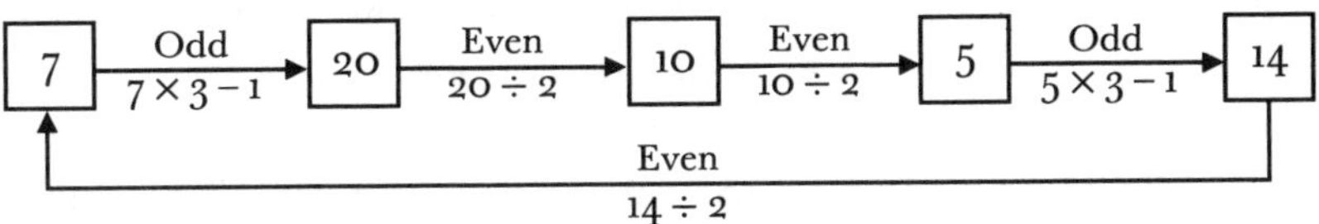

This is a hailstone that gets caught in an endless cycle of updrafts and downdrafts but ends up back where it started so that it never hits the ground, which for us is the number 1. On the other hand, if we replaced the rule for odd numbers with "multiply by 3 and add 2," we would end up with a very different result. If we start with the number 12, we get the following sequence.

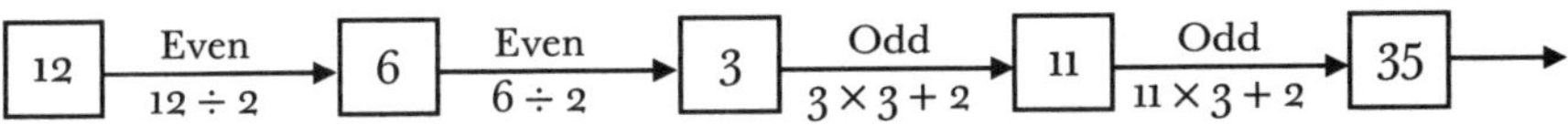

I will leave it to you to continue this sequence. The problem here is that once the number is odd, it will continue to be odd. After all, 3 times an odd number is odd, and adding 2 to it keeps it odd. As such, the hailstone number has escaped mathematical gravity altogether and is now on its way to outer space. The only way to reach 1 is first to get to a power of 2, and in this modification of the Collatz process that can only happen by starting on a power of 2. Every other number eventually will go off to infinity.

Therefore, we have two variations on the Collatz Conjecture, one that has examples that end in infinite loops and one that has examples that skyrocket off to infinity. This brings us back to the original question. Is the Collatz Conjecture true? We simply do not know. We have not found any numbers that do *not* eventually end at 1, but we cannot prove that it *always* happens. Some of the sequences that we have found are quite long. For example, even a relatively small number like 27, which eventually does reach 1, takes 111 steps to get there!

Paul Erdős said of this conjecture, "Mathematics is not yet ripe enough for such questions." One of the reasons it may be so difficult is the seemingly random sequences of numbers it produces. Other than the pattern that appears simply because it comes from a known process, the numbers seem to go up and down unpredictably. Why bother working on this problem, with its playful nature? One answer is that solving the Collatz Conjecture could give us a deeper insight into how "easy" it is to avoid powers of 2. A second answer could be that new, deep and wonderful mathematics is needed to solve this problem, and the mere development of those techniques could lead to other advances in the discipline. This was certainly the case with Fermat's Last Theorem. As wonderful as it was to see the conquering of a problem casually mentioned in the margins of a lawyer's copy of the *Arithmetica*, there is not a mathematician alive who does not acknowledge the importance of the mathematics surrounding its solution. It is not too much to say that the quest to solve Fermat's Last Theorem, with its many contributors, produced entire *fields* of mathematics that are now rich with interesting results and arrays of beautiful open questions remaining to be solved. Finally, there is the mere curiosity of the Collatz Conjecture itself, which is made all the more compelling by its apparent difficulty. After all, how can a problem that is so easy to state be so difficult to solve?[192]

192 This is a truly marvelous open question that can be introduced to elementary school students. I alluded to this at the start of the chapter. Young students can perform the Collatz calculations and explore the results, which strengthens both their arithmetic skills and their understanding of even numbers, odd numbers, and multiples of 3. On top of all that, it is fun to be able to tell them, "It seems like all numbers eventually land on 1, but *nobody knows whether that is always the case!* Maybe *you* can find one that does not!" For the love of all that is good, however, do not tell them that computers have checked very large numbers and failed to find any counterexamples. Let the students explore with the optimism that they may make a discovery that nobody else has made.

CHAPTER 8
THE 196-ALGORITHM

LIKE THE Collatz Conjecture, this problem has a bit of a recreational flare to it. Start by taking any positive number, for example 38. First, reverse the digits to get 83, then add the two numbers together.

$$\begin{array}{r} 38 \\ +\ 83 \\ \hline 121 \end{array}$$

The result is a palindrome, meaning that it reads the same forwards as backwards. This alone may not be very interesting. After all, it doesn't happen with every number. Take, for example, the number 67.

$$\begin{array}{r} 67 \\ +\ 76 \\ \hline 143 \end{array}$$

The result is *not* a palindrome. But what if we continue the process? Reverse the digits of 143 and add the two numbers together.

$$\begin{array}{r} 143 \\ +341 \\ \hline 484 \end{array}$$

This sum *is* a palindrome; it just took two steps instead of one to get there. The amazing thing is that this seems to happen *a lot*, even though it takes a few more steps for some numbers to arrive at their palindromic plentitude. Consider starting with 96.

$$\begin{array}{ll} 96+69 & = 165 \\ 165+561 & = 726 \\ 726+627 & = 1{,}353 \\ 1{,}353 + 3{,}531 & = 4{,}884 \end{array}$$

In this case, the result is a palindrome after four steps. While certainly more than one or two steps, four steps turns out to be fairly tame. After all, the number 10,911 takes fifty-five steps before it reaches a twenty-eight-digit palindrome, yet it gets there eventually. *Does this happen for every number?* With the Collatz Conjecture, we saw that every number seems eventually to end its Collatz process at 1. Mathematicians and computer scientists have checked the Collatz Conjecture for very large values, and thus far every number does indeed end at 1. This is precisely why mathematicians suspect that the Collatz Conjecture is true, even though they cannot find a proof. The case of the palindrome-seeking process is quite different. Mathematicians are fairly certain that the answer is no. The conjecture is that *not* every number will eventually end in a palindrome, and the reason why is the pesky number 196.

196 + 691	= 887
887 + 788	= 1,675
1,675 + 5,761	= 7,436
7,436 + 6,347	= 13,783
13,783 + 38,731	= 52,514
52,514 + 41,525	= 94,039
94,039 + 93,049	= 187,088
187,088 + 880,781	= 1,067,869

This number seems never to reach a palindrome. Lest we think that perhaps we have not gone far enough—after all, we just mentioned that 10,911 takes fifty-five steps to become palindromic—we should note that mathematicians have continued the reverse-and-add process for 196 to produce some *very* large numbers. So far, it still never seems to reach a palindromic state. To be fair, this number is not alone. There are other "problem" numbers, including 295, 394, 493, and 592. There actually may be an infinite number of these problematic non-palindrome-producing numbers. Such numbers are known as *Lychrel numbers*, a term coined by the computer scientist Wade Van Landingham.[193]

The conjecture is simply that Lychrel numbers actually exist.

> *The Existence of Lychrel Numbers Conjecture*
> There are some numbers, called *Lychrel numbers*,
> that will never end in a palindrome when sent through the
> "reverse the digits and add the results" algorithm.
> Specifically, the first Lychrel number is 196.

193 Landingham himself has said that the term "Lychrel" has no meaning except that it was a rough anagram of his girlfriend's name, that the term came to him while driving, and that he "liked the sound of it."

Like the Collatz Conjecture, this is another one of those great algorithm-driven problems to give to young students. Some examples will end easily after a small number of steps, but others will prove to be quite challenging. The number 89 might not be as bad as the aforementioned 10,911, but given how tame the number seems, it is a bit surprising that it takes twenty-four steps to reach a thirteen-digit-long palindromic bliss. Just because 196 and other numbers seem to take a while, why would we suspect that they *never* reach a palindromic state? One reason is what happens in different bases.

As early as first and second grade, we learn to count and write numbers by thinking about groups of 1, groups of 10, groups of 100, groups of 1,000, etc. The number 6,782 literally means "6 thousands, 7 hundreds, 8 tens, and 2 ones." It also means $6\times1{,}000+7\times100+8\times10+2\times1$, or $6\times10^3+7\times10^2+8\times10^1+2\times1$. The place values are powers of ten. One great thing about place value notation is that it allows us to express every positive whole number in exactly one way. It also paves the way for the beautiful standard algorithms we learn for addition, subtraction, multiplication, and even long division. (Try doing long division using Roman numerals, and you will come to appreciate base ten notation.)

That said, there is no particular reason why we need ten symbols (0, 1, 2, 3, 4, 5, 6, 7, 8, 9) for our system.[194] What if we only had *two* symbols: 0 and 1? In this "base two" system, more commonly known as *binary*, the place values are not powers of ten but powers of two. For example, the number 11011 in binary means $1\times2^4+1\times2^3+0\times2^2+1\times2^1+1$. We can convert this number to base ten by multiplying it all out.

$$\begin{aligned} 1\times2^4+1\times2^3+0\times2^2+1\times2^1+1 &= 1\times16+1\times8+0\times4+1\times2+1\times1 \\ &= 16+8+0+2+1 \\ &= 27 \end{aligned}$$

It is tempting to present an entire chapter just on the beauty of the place value system, the particulars of the binary system, and why the standard algorithms in different bases still work. Such a tangent, however, could distract us from the main question at hand, which is the existence of

194 Although we could have chosen any base to use when writing and calculating with numbers, the fact that we have ten fingers is precisely why we ended up with a system that organizes numbers into groups of ten.

Lychrel numbers. The reason why the discussion of other bases is relevant is that we *know* that the number 10110 is a Lychrel number in the binary system. The existence of binary Lychel numbers is not a conjecture; it is a well-established theorem.

> *Lychrel Numbers in Binary Theorem*
> Lychrel numbers *do* exist in the binary system.
> Specifically, the number 10110 is a binary Lychrel number.

Wait—if numbers can be converted back and forth between base ten and base two, wouldn't it make sense that if a number is a Lychrel number in base two, then it should also be a Lychrel number in base ten? In other words, isn't it true that "once Lychrel, always Lychrel"? After all, "twenty-seven apples" is the same number of apples whether the number is written as 27 in base ten or as 11011 in binary. The problem is that the property of being a palindrome does not cross over from one base to another. In fact, the exact example of 27 and 11011 demonstrates this non-transferability of palindromic peacefulness. The number 27 is not a palindrome, but when it is written in base two, 11011 *is* a palindrome.

Getting back to the theorem, how did mathematicians prove that the binary number 10110 will never arrive at a palindrome, and why can a similar proof not be used to show that 196 is a Lychrel number in base ten? The way mathematicians proved the "Lychrelity" of the binary number 10110 includes two steps.

1. There is an observable pattern for the results at steps 4, 8, 12, 16, etc. The number always starts with a 10 and is followed by a certain number of 1s, then a 01, and finally the same number of 0s that were in the string of 1s. Mathematicians noticed this pattern and proved that it always happens. Because the numbers at these steps start with a 1 and end with a 0, they are never palindromes.

2. They then showed that the steps between these predictable steps (1, 2, 3, 5, 6, 7, 9, 10, 11, ...) can never produce palindromes, either.

A similar strategy might be used to show that 196 is a Lychrel number *if* we could find such a pattern in its steps. To date, no such pattern has been detected. The big takeaway here is that Lychrel numbers do exist, at least in base two. Moreover, they are known to exist in all bases that are powers of 2: 2, 4, 8, 16, 32, 64, 128, etc. Then, as if randomly, we have also found them in base 11, 17, 20, and 26. Therefore, if Lychrel numbers exist in *some* bases, and if a number like 196 *seems* to be a Lychrel number in base ten, the reasonable conjecture is that it is indeed Lychrel. Yet, as with all unsolved problems, we do not know for sure. The 196-algorithm remains a mystery. Will it ever produce a palindrome, or is it an authentic Lychrel number?

CHAPTER 9
IS $\pi + e$ A RATIONAL NUMBER? HOW ABOUT $\pi \times e$?

WHEN WE looked at irrational numbers, we talked about the basic arithmetic operations and how there are some numbers, namely $\sqrt{2}$, that cannot be written as a division problem involving integers. In other words, there is no fraction of integers that equals $\sqrt{2}$. Irrational numbers seem to be "wholly other" than rational numbers. We can start with any rational number we want and combine it with other rational numbers using addition, subtraction, multiplication, and division, and we will *never* get an irrational number. And yet, as we saw in a later chapter, there are a lot more irrational numbers than rational numbers even though most of us only know about a handful of them.

It isn't just that the irrational numbers seem out of reach of the rational ones along the byways of simple arithmetic. Rather, it is that once an irrational number is introduced into an operation, the result seems to stay "outside" the tidy world of rational numbers. For example, adding an irrational number to a rational number will always produce an irrational number. The rational number has been "tainted" by having a bit of irrationality added to it. Let's consider a specific example: $\frac{2}{3}+\sqrt{2}$. I know this is an addition problem, but let's think for a moment about subtraction. We already know that the rational numbers are closed under subtraction. In other words, a rational number subtracted from another rational number is always a rational number. If $\frac{2}{3}+\sqrt{2}$ were a rational number, then the difference between $\frac{2}{3}+\sqrt{2}$ and $\frac{2}{3}$ would also be rational. The difference, however, is $\left(\frac{2}{3}+\sqrt{2}\right)-\frac{2}{3}=\sqrt{2}$. In other words, if $\frac{2}{3}+\sqrt{2}$ were rational, then $\sqrt{2}$ would also be rational. But we went to great lengths to prove that $\sqrt{2}$ is *not* rational. Therefore, $\frac{2}{3}+\sqrt{2}$ cannot possibly be rational.

The above argument can be generalized to show that the sum of a rational number and an irrational number will always be irrational. A virtually identical argument can be made to show that the *difference* between a rational and an irrational number is always irrational. (Can you work out the details of that argument to show that $\frac{7}{11}-\sqrt{2}$ must be irrational?)

What about the product of an irrational number and a rational number, for example $\frac{2}{3}\times\sqrt{2}$? We already know that the result of multiplying two rational numbers is always rational. Again, the rational numbers are *closed* under multiplication. If $\frac{2}{3}\times\sqrt{2}$ were rational, then the product $\frac{3}{2}\times\left(\frac{2}{3}\times\sqrt{2}\right)$ would also be rational. This product, however, is $\frac{3}{2}\times\left(\frac{2}{3}\times\sqrt{2}\right)=\sqrt{2}$. In other words, if $\frac{2}{3}\times\sqrt{2}$ is rational, so is $\sqrt{2}$. Therefore, $\frac{2}{3}\times\sqrt{2}$ must be irrational. As before, the choice of $\frac{2}{3}$ and $\sqrt{2}$ was fairly arbitrary. We can generalize the argument to show that the product of a rational number and an irrational number is always irrational, except for the one case when the rational number is 0. A similar argument can be made to show that the quotient of a rational number and an irrational number is always irrational, so long as the rational number is not 0.

Addition	Subtraction
Rational + Rational = Rational	Rational − Rational = Rational
Rational + Irrational = Irrational	Rational − Irrational = Irrational Irrational − Rational = Irrational
Multiplication	**Division**
Rational × Rational = Rational	Rational ÷ Rational = Rational (so long as we are not dividing by 0)
Rational × Irrational = Irrational (so long as the rational is not 0)	Rational ÷ Irrational = Irrational Irrational ÷ Rational = Irrational (so long as the rational is not 0)

The rational numbers form a tidy world. They behave perfectly well on their own and in some ways have no need for the irrational numbers. Moreover, introducing an irrational number into the mix using the four basic operations will taint the original rational number and produce an irrational result (except for those pesky cases involving 0). It seems like once we introduce an irrational number, we leave rationality behind and enter the strange world of irrationality. Is this really the case? Are we really "stuck" once we enter the irrational numbers? Is there no path back into the tidy, self-contained world of rationality? While it is true that no irrational number can be the result of adding, subtracting, multiplying, or dividing rational numbers, it is perhaps surprising to find out that the reverse is decidedly not the case.

It turns out that we can take two *irrational* numbers, add them together, and find the result to be a *rational* number. Consider the sum of the following two numbers.

$$2.080880229\ldots + 1.252453104\ldots$$

For just a moment, let's assume that the decimals for each number do not repeat and do not terminate. That is, let's trust that these are indeed two irrational numbers. When we add these two numbers together, we get $3.333333333\cdots = \frac{10}{3}$, which is perfectly rational! When we look more closely at the two numbers being added together, it might not be all that surprising. It turns out that the original two numbers are $\frac{2}{3}+\sqrt{2}$ and $\frac{8}{3}-\sqrt{2}$. First note that both numbers are irrational. We just proved that fact. The sum of these two numbers, however, is $\left(\frac{2}{3}+\sqrt{2}\right)+\left(\frac{8}{3}-\sqrt{2}\right)=\left(\frac{2}{3}+\frac{8}{3}\right)+\left(\sqrt{2}-\sqrt{2}\right)=\frac{10}{3}$, which is rational. I will leave it to you to use a similar technique to show that we can subtract two irrational numbers and end up with a rational result. Now, to be sure, *sometimes* the sum or difference of two irrational numbers is indeed irrational, as in the case of $\sqrt{2}+\sqrt{3}$. It is just that this is not *always* the case.

What about multiplication? Surely it is not possible to multiply two things that cannot be written as a fraction of integers and get a result that *is* a fraction of integers. Well, consider the following product.

$$\left(\frac{\sqrt{2}}{5}\right)\times\left(\frac{\sqrt{2}}{3}\right)$$

As the quotient of an irrational number and a rational number, $\frac{\sqrt{2}}{5}$ and $\frac{\sqrt{2}}{3}$ are irrational per our proof above. Yet the product is $\left(\frac{\sqrt{2}}{5}\right)\times\left(\frac{\sqrt{2}}{3}\right)=\frac{\sqrt{4}}{15}=\frac{2}{15}$, which is a rational number. So even the *product* of two irrational numbers can be rational. I will leave it to you to discover two irrational numbers whose quotient is rational. Again, to be sure, *sometimes* the product or quotient of two irrational numbers is irrational, as in the case of $\sqrt{2}\times\sqrt{3}$. It is just that this is not *always* the case.

There are actually much simpler examples of these phenomena: $\pi+(-\pi)=0$, $\pi+(1-\pi)=1$, $\pi+(2-\pi)=2$, and so on. Similarly, $\pi\times\frac{1}{\pi}=1$, $\pi\times\frac{2}{\pi}=2$, $\pi\times\frac{3}{\pi}=3$, and so on. In fact, these examples show that the sums and products of transcendental numbers can be rational, which is perhaps more amazing given that the transcendental numbers *transcend* the basic arithmetic operations. And yet, these examples are not mind-blowing. After all, the first set relies on subtracting π from itself and the second on dividing π by itself. Instances like this—where the problem first seems incredibly complicated, but the solution is simple—often elicit the reaction, "That was obvious! How could I not see that in the first place?" When this happens, it is hard to forget that singular moment of clarity, something akin to our apprehension of beauty. Even though beauty might have always been present without our notice, a moment of insight happens, and we *see*.

It is worth adding to our chart to summarize how rational numbers and irrational numbers play together within the four basic operations.

Addition	Subtraction
Rational + Rational = Rational	Rational − Rational = Rational
Rational + Irrational = Irrational	Rational − Irrational = Irrational Irrational − Rational = Irrational
Irrational + Irrational = Rational or Irrational	Irrational − Irrational = Rational or Irrational
Multiplication	**Division**
Rational × Rational = Rational	Rational ÷ Rational = Rational (so long as we are not dividing by 0)
Rational × Irrational = Irrational (so long as the rational is not 0)	Rational ÷ Irrational = Irrational Irrational ÷ Rational = Irrational (so long as the rational is not 0)
Irrational × Irrational = Rational or Irrational	Irrational ÷ Irrational = Rational or Irrational

One of the more surprising combinations happens with irrational exponents. It is not a surprise that an irrational number raised to an irrational power normally results in another irrational number. It turns out, however, that sometimes an irrational number raised to an irrational power can produce a rational result. The proof of this is an exceptionally elegant *existence proof*. An existence proof shows that something must exist without actually producing a definitive example of it.

In this case, we start with the number $\sqrt{2}^{\sqrt{2}}$. We might not know whether this number is rational or irrational, but it must be one of the two. Of course, if it is rational, then we have completed our task. We have found an irrational number, $\sqrt{2}$, raised it to an irrational power, $\sqrt{2}$, and obtained a rational result.

What if it is not rational? What if $\sqrt{2}^{\sqrt{2}}$ is an *irrational* number? The brilliant move is to raise this number itself to the power of $\sqrt{2}$.

$$\left(\sqrt{2}^{\sqrt{2}}\right)^{\sqrt{2}}$$

If $\sqrt{2}^{\sqrt{2}}$ is irrational, then we have an irrational number raised to an irrational power, $\sqrt{2}$. But when we simplify our number, something magnificent happens.

$$\left(\sqrt{2}^{\sqrt{2}}\right)^{\sqrt{2}} = \sqrt{2}^{\sqrt{2} \times \sqrt{2}} = \sqrt{2}^{2} = 2$$

The number 2 is a rational number. We now have an irrational number raised to an irrational power with a result that is rational.

The amazing thing about this proof is that we never actually determined whether $\sqrt{2}^{\sqrt{2}}$ is rational or irrational. In other words, we never found an explicit example of an irrational number raised to an irrational power being rational. What we *did* discover is that such an example must be either $\sqrt{2}^{\sqrt{2}}$ or $\left(\sqrt{2}^{\sqrt{2}}\right)^{\sqrt{2}}$. We just don't know which one fits the bill.[195]

Given how much we know about the interplay between rational and irrational numbers, it is perhaps all the more surprising that we do not know whether the sum of the world's most famous irrational numbers, e and π, is rational or irrational. The same is true concerning their product.

> UNSOLVED PROBLEM: Are the numbers $\pi + e$ and $\pi \times e$ rational or irrational?

These two situations are not alone: there are several number combinations for which we do not know whether the result is rational or irrational. A partial list includes $\pi + e$, $\pi - e$, $\pi \times e$, $\pi \div e$, π^{e}, $\pi^{\sqrt{2}}$, π^{π}, e^{e}, and 2^{e}.

195 Actually, mathematicians do know whether $\sqrt{2}^{\sqrt{2}}$ is rational or irrational, but I will not spoil the result for the reader.

CHAPTER 10
A QUESTION ABOUT FRIENDS AND STRANGERS

SIX PEOPLE attend a party. We will call them Christina, Tatiana, Kateri, Jude, Benedict, and Sebastian. Some of them are meeting for the first time, and others knew each other before coming to the party. In the first case we will call them strangers, and in the second case we will call them friends. I claim that there must either be three mutual friends or three mutual strangers. For example, consider a vertex-edge graph in which a solid line means that the two people are friends.

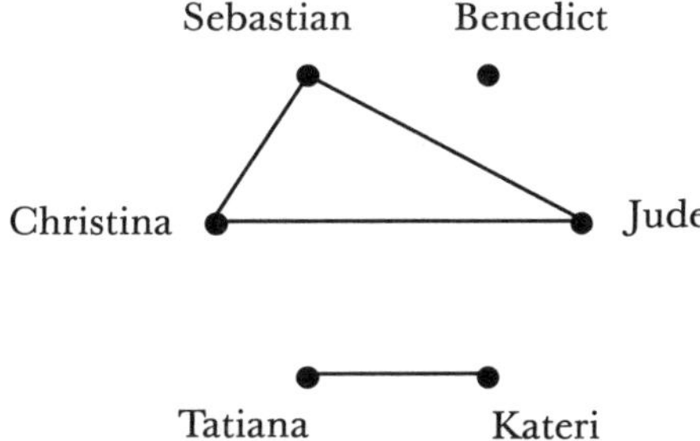

Jude and Sebastian are friends, but Christina and Kateri are strangers. In this case, there are three mutual friends: Sebastian, Christina, and Jude.

This is not always the case, though. Consider another possibility.

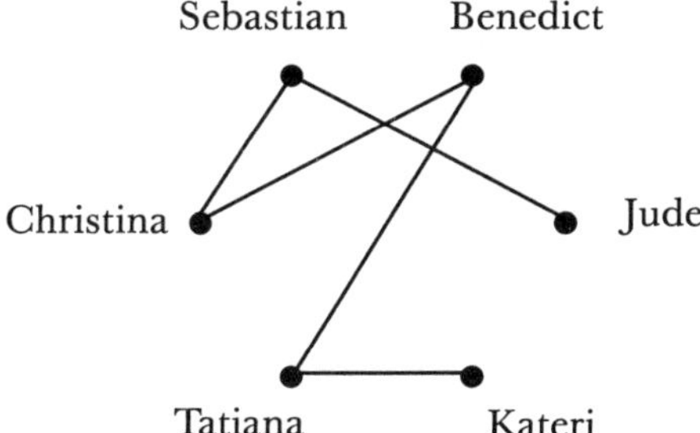

In this case, there are not three mutual friends. There are, however, three mutual strangers: Sebastian, Benedict, and Kateri. The three mutual friends in the first example were easier to see in the diagram because a triangle was formed by the three names. When there are three mutual strangers, there is an absence of a triangle altogether, meaning that the three names—Sebastian, Benedict, and Kateri—do not have any lines in the would-be triangle. It might be easier to see the relationship of strangers if instead of an absence of a line, we used a different kind of line.[196]

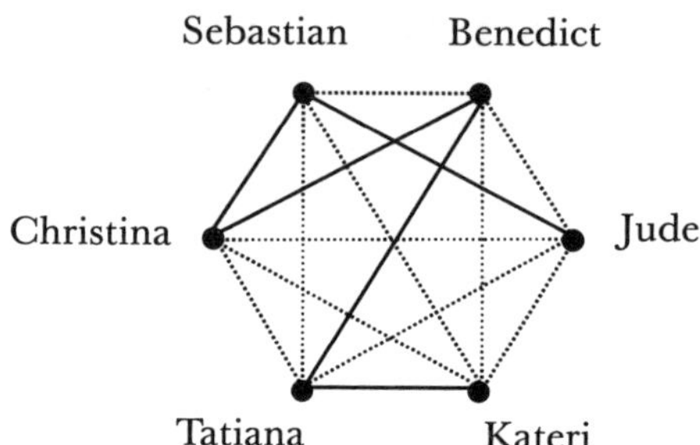

We can now see that the Sebastian–Benedict–Kateri triangle is formed by three dotted lines, indicating that they are three mutual strangers.

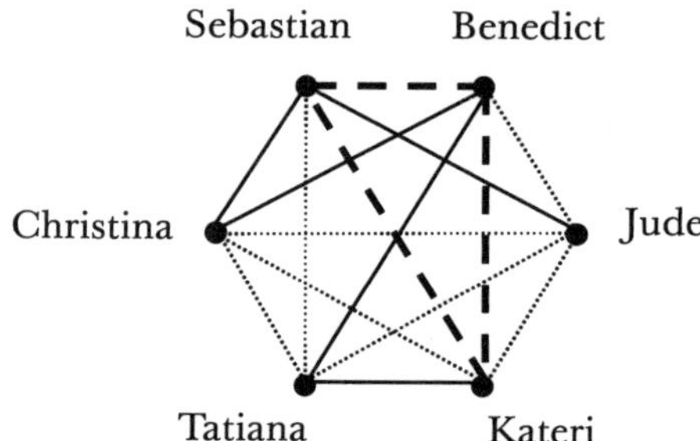

We can also see two other examples of three mutual strangers. Can you find them?

These two examples do not prove my claim. They only show that there *could be* three mutual friends or three mutual strangers in this party of six. My claim was considerably stronger. I insisted that there *must be* either three mutual strangers or three mutual friends. In other words, no matter how you draw the diagram, there will either be a solid-line triangle or a

196 I like the dotted line better than the absence of a line for another reason: it suggests that someday Benedict and Jude might be friends!

dotted-line triangle. You cannot avoid both. In fact, this claim of "must" is more interesting when you consider that it does not hold for a party of five. Consider a different set of guests.

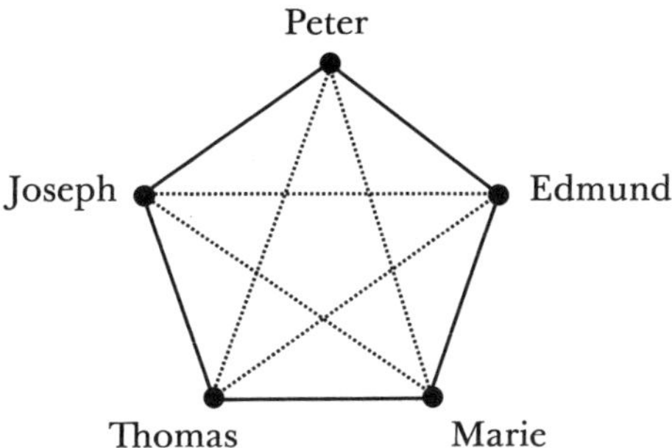

There are no solid-line triangles and no dotted-line triangles. Therefore, there is neither a case of three mutual friends nor a case of three mutual strangers. This counterexample shows that with five people at the party, the property is not guaranteed. Something happens when we move to six people that forces us into one or the other case.

To help us see why, let's look at a blank diagram with our six original guests.

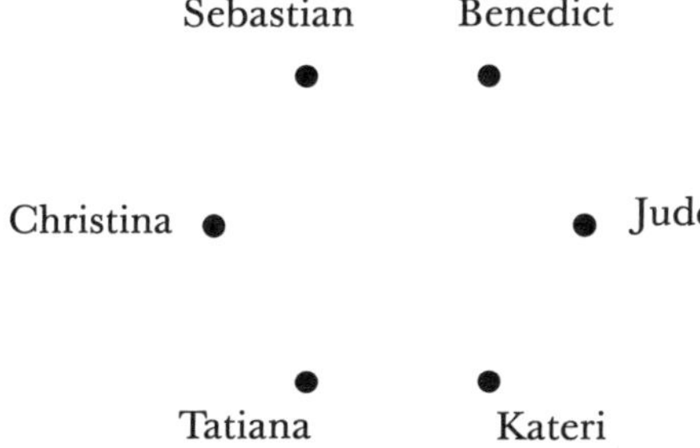

Notice that in a completed diagram every possible pair of guests will have a line, either dotted or solid, because any two guests are either friends or strangers. This means that there must be exactly five lines coming from Christina, one to each of the other guests. Some will be dotted, and others will be solid. (You can count them in the second example. There were three dotted lines and two solid lines.) For Christina's five lines, there are six possibilities.

0 dotted, 5 solid
1 dotted, 4 solid
2 dotted, 3 solid
3 dotted, 2 solid
4 dotted, 1 solid
5 dotted, 0 solid

Notice that in all cases there are at least three solid lines or at least three dotted lines. Sometimes there are more. For example, in the first case there are five solid lines, which certainly satisfies "at least three." Therefore, Christina must have at least three friends or at least three strangers. Let's consider each case separately.

CASE 1: *Christina has at least three friends.*

Suppose that there are at least three solid lines coming off of Christina. Where do those lines go? We are not sure, but let's call them persons *X*, *Y*, and *Z*. This means that Christina is friends with all three of these people. We will draw them apart from the hexagon so that you do not think that *X*, *Y*, and *Z* are specific people. They could be any of the five people who are not Christina.

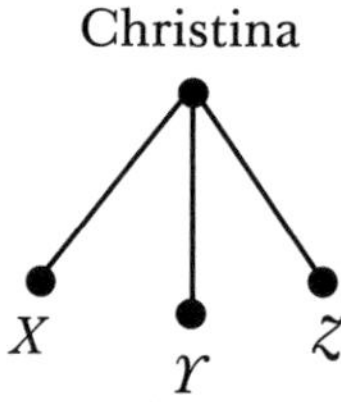

If *any* of *X*, *Y*, and *Z* are also friends, then we will have a triangle of three mutual friends. For example, if *Y* and *Z* are friends, then we have a solid-line triangle consisting of three mutual friends, one of whom is Christina.

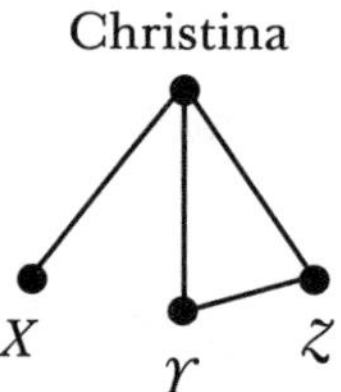

But if *none* of X, Y, and Z are friends, then we have a dotted-line triangle formed by three mutual strangers in X, Y, and Z.

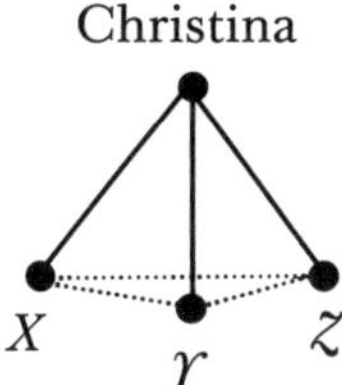

The result is that there are either three mutual friends or three mutual strangers.

CASE 2: *Christina has at least three strangers.*

The argument is identical, and I will leave it to you to draw the accompanying diagrams. Suppose that there are at least three dotted lines coming off of Christina. Where do those lines go? Again, we are not sure, but let's call them persons X, Y, and Z. This means that Christina is a stranger to all three of these people. If *any* of X, Y, and Z are also strangers, then we will have a triangle of three mutual strangers. But if *none* of X, Y, and Z are strangers, then they themselves form a triangle of three mutual friends.[197]

In any case, no matter how we slice it, there will either be three mutual friends or three mutual strangers.

> *The Six-Person Party Pre-Theorem*
> Whenever there are six people at a party, there must be three people that are either mutual friends or mutual strangers.

Of course, with seven people, there also must be either three mutual friends or three mutual strangers. We could simple pick any six of the seven and then apply the Six-Person Party Pre-Theorem to them. Moreover, as we saw, the claim does not necessarily hold for *five* people. That means that we have a somewhat stronger theorem.

197 While not necessary for the validity of the argument, it is worth asking why this same strategy does not apply to the case of a party of five, for which we already know that the property does not hold. It would be troubling indeed if this proof "worked" for something we know to be false. The argument breaks down in the very first step. In the case of a party of five, there are only four lines coming off of Peter. With only four lines, there is no guarantee that there are at least three dotted lines or three solid lines, which was key for the proof. Rather, there could be exactly two dotted lines and two solid lines, and in fact that is exactly the case in the pentagonal counterexample. Without this guarantee, the rest of the argument falls apart.

The Six-Person Party Theorem
The first party size that guarantees either three mutual friends or three mutual strangers is six.

We might ask, "How many people do we need at the party in order to guarantee that there are either *four* mutual friends or *four* mutual strangers?" This problem is much more cumbersome to solve, but the answer is known.

The Eighteen-Person Party Theorem
The first party size that guarantees either four mutual friends or four mutual strangers is eighteen.

Not content with the previous two theorems, a mathematician might ask, "How many people do we need at the party in order to guarantee that there are either *five* mutual friends or *five* mutual strangers?" The answer to this question, as it turns out, is unknown.

UNSOLVED PROBLEM: What is the first party size that guarantees either five mutual friends or five mutual strangers?

There has been some work on this problem, and we do know two important things.

1. With forty-two people at the party, we can *avoid* having five mutual friends and five mutual strangers.
2. With forty-eight people at the party, there *must* be either five mutual friends or five mutual strangers.

This means that the answer to the unsolved problem is either 43, 44, 45, 46, 47, or 48. It should be noted that lots of examples of forty-two people have been found that avoid having five mutual friends and five mutual strangers, and with no examples found for forty-three, the current conjecture is that the answer to the unsolved problem is forty-three.[198]

198 These numbers are known as Ramsey numbers, named after the British mathematician F. P. Ramsey. Ramsey numbers are a bit more general, though. They allow for the number of friends to differ from the number of strangers. For example, R(3, 3) is defined as the smallest number of people necessary in order to guarantee either three mutual

We could think about solving this problem by writing a computer program to generate all possibilities for forty-three people. The problem is that this program would run for longer than the time period most of humanity would be alive. There are 2^{903} possible situations, a staggering number. We can do some clever elimination of cases based on symmetry, but the number of possibilities will still be far more than the number of particles in the known universe. That said, Paul Erdős sees at least a glimmer of hope in solving this problem. He imagines alien life forms threatening to destroy Earth if we cannot answer their question: "What is the least number of people at a party that guarantees either five mutual friends or five mutual strangers?" In this case, Erdős says we should "marshal all our computers and all our mathematicians and attempt to find the value."

With that, we have reached the end of our journey and have come full circle to at least one of our starting points. Mathematics and mathematicians proceed under the principle of *contra utilitatem solam*, or "against utility alone." Several mathematical giants—a strange number of whom have a name that starts with "E"—have been prominent along the way, from Euclid to Euler. The honor of having something of the last word, however, goes to Erdős. While mathematics proceeds not by utility alone, at least in this one case, knowing the answer to our final unsolved problem could save us from alien domination, a useful thing indeed.

Lest we become too comfortable, what if instead the aliens ask, "What is the least number of people at a party that guarantees either *six* mutual friends or *six* mutual strangers?" Erdős believes that the task would be nearly impossible for even the fastest of computers and that in the face of such a threat "we should attempt to destroy the aliens."[199]

---

friends or three mutual strangers, which we now know to be six, but R(3, 4) represents the smallest number of people necessary in order to guarantee either three mutual friends or four mutual strangers. The unsolved problem is to find the value of R(5, 5). It is also the case that Ramsey numbers are defined in terms of colored lines, akin to our dotted and solid lines, on a vertex-edge graph rather than in terms of friends and strangers.

199 This delightful attribution to Erdős appears on page 4 of Joel Spencer's book *Ten Lectures on the Probabilistic Method* (Society for Industrial and Applied Mathematics, 1994). The quotation uses the notation for Ramsey numbers rather than "friends and strangers," but the content is otherwise the same.

EPILOGUE

At the end of the introduction, I issued a charge worth repeating: embrace the mathematician within you; choose to wonder. I also humbly noted that most of the ideas contained in this book are not my own. They are the result of greater minds who have come before me or in some cases continue to accompany me. It seems only fitting, then, that I give some of them the last words.

> *Why are numbers beautiful? It's like asking why Beethoven's Ninth Symphony is beautiful. If you don't see why, someone can't tell you.*
>
> —PAUL ERDŐS

> *There is something unique in the human soul that can only be satisfied by wondering about mathematics.*
>
> —MICHAEL AUSTIN

$$e^{\pi i}+1=0.$$

—GOD

INDEX OF THEOREMS AND LEMMAS

INDEX OF UNSOLVED PROBLEMS

INDEX

[E]

[P]

[R]